EARTHQUAKE ENGINEERING
FOR STRUCTURAL DESIGN

EARTHQUAKE ENGINEERING FOR STRUCTURAL DESIGN

EDITED BY
W.F. Chen
E.M. Lui

Taylor & Francis
Taylor & Francis Group
Boca Raton London New York

A CRC title, part of the Taylor & Francis imprint, a member of the
Taylor & Francis Group, the academic division of T&F Informa plc.

This material was previously published in *Handbook of Structural Engineering, Second Edition.* © CRC Press LLC 2005

Published in 2006 by
CRC Press
Taylor & Francis Group
6000 Broken Sound Parkway NW, Suite 300
Boca Raton, FL 33487-2742

International Standard Book Number-10: 0-8493-7234-8 (Hardcover)
International Standard Book Number-13: 978-0-8493-7234-6 (Hardcover)
Library of Congress Card Number 2005050642

Library of Congress Cataloging-in-Publication Data

Earthquake engineering for structural design / Wai-Fah Chen, Eric M. Lui [editors].
    p. cm.
Includes bibliographical references and index.
ISBN 0-8493-7234-8 (alk. paper)
    1. Earthquake engineering. 2. Structural design. I. Chen, Wai-Fah, 1936- II. Lui, E. M.

TA654.6.E372 2005
624.1'762--dc22   2005050642

is the Academic Division of Informa plc.

Visit the Taylor & Francis Web site at
http://www.taylorandfrancis.com

and the CRC Press Web site at
http://www.crcpress.com

The Editors

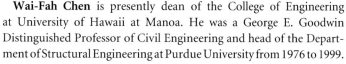

peng chen #1

Wai-Fah Chen is presently dean of the College of Engineering at University of Hawaii at Manoa. He was a George E. Goodwin Distinguished Professor of Civil Engineering and head of the Department of Structural Engineering at Purdue University from 1976 to 1999.

He received his B.S. in civil engineering from the National Cheng-Kung University, Taiwan, in 1959, M.S. in structural engineering from Lehigh University, Pennsylvania, in 1963, and Ph.D. in solid mechanics from Brown University, Rhode Island, in 1966.

Dr. Chen received the Distinguished Alumnus Award from National Cheng-Kung University in 1988 and the Distinguished Engineering Alumnus Medal from Brown University in 1999.

Dr. Chen is the recipient of numerous national engineering awards. Most notably, he was elected to the U.S. National Academy of Engineering in 1995, was awarded the Honorary Membership in the American Society of Civil Engineers in 1997, and was elected to the Academia Sinica (National Academy of Science) in Taiwan in 1998.

A widely respected author, Dr. Chen has authored and coauthored more than 20 engineering books and 500 technical papers. He currently serves on the editorial boards of more than 10 technical journals. He has been listed in more than 30 *Who's Who* publications.

Dr. Chen is the editor-in-chief for the popular 1995 *Civil Engineering Handbook*, the 1997 *Structural Engineering Handbook*, the 1999 *Bridge Engineering Handbook*, and the 2002 *Earthquake Engineering Handbook*. He currently serves as the consulting editor for the McGraw-Hill's *Encyclopedia of Science and Technology*.

He has worked as a consultant for Exxon Production Research on offshore structures, for Skidmore, Owings and Merrill in Chicago on tall steel buildings, for the World Bank on the Chinese University Development Projects, and for many other groups.

peng chen #2

Eric M. Lui is currently chair of the Department of Civil and Environmental Engineering at Syracuse University. He received his B.S. in civil and environmental engineering with high honors from the University of Wisconsin at Madison in 1980 and his M.S. and Ph.D. in civil engineering (majoring in structural engineering) from Purdue University, Indiana, in 1982 and 1985, respectively.

Dr. Lui's research interests are in the areas of structural stability, structural dynamics, structural materials, numerical modeling, engineering computations, and computer-aided analysis and design of building and bridge structures. He has authored and coauthored numerous journal papers, conference proceedings, special publications, and research reports in these areas. He is also a contributing author to a number of engineering monographs and handbooks, and is the coauthor of two books on the subject of structural stability. In addition to conducting research, Dr. Lui teaches a variety of undergraduate and graduate courses at Syracuse University. He was a recipient of the College of Engineering and Computer Science Crouse Hinds Award for Excellence in Teaching in 1997. Furthermore, he has served as the faculty advisor of Syracuse University's chapter of the American Society of Civil Engineers (ASCE) for more than a decade and was recipient of the ASCE Faculty Advisor Reward Program from 2001 to 2003.

Dr. Lui has been a longtime member of the ASCE and has served on a number of ASCE publication, technical, and educational committees. He was the associate editor (from 1994 to 1997) and later the book editor (from 1997 to 2000) for the ASCE *Journal of Structural Engineering*. He is also a member of many other professional organizations such as the American Institute of Steel Construction, American Concrete Institute, American Society of Engineering Education, American Academy of Mechanics, and Sigma Xi.

He has been listed in more than 10 *Who's Who* publications and has served as a consultant for a number of state and local engineering firms.

Contributors

Wai-Fah Chen
College of Engineering
University of Hawaii at Manoa
Honolulu, Hawaii

Lian Duan
Division of Engineering Services
California Department of Transportation
Sacramento, California

Ronald O. Hamburger
Simpson Gumpertz & Heger, Inc.
San Francisco, California

Sashi K. Kunnath
Department of Civil and Environmental
  Engineering
University of California
Davis, California

Mark Reno
Quincy Engineering
Sacramento, California

Charles Scawthorn
Department of Urban Management
Kyoto University
Kyoto, Japan

Shigeki Unjoh
Ministry of Construction
Public Works Research Institute
Tsukuba, Ibaraki, Japan

Mark Yashinsky
Division of Structures Design
California Department of Transportation
Sacramento, California

Contents

1

Fundamentals of Earthquake Engineering

Charles Scawthorn
Department of Urban Management,
Kyoto University,
Kyoto, Japan

1.1 Introduction

This chapter provides a basic understanding of earthquakes, by first discussing the causes of earthquakes, then defining commonly used terms, explaining how earthquakes are measured, discussing the distribution of seismicity, and, finally, explaining how seismicity can be characterized.

Earthquakes are broad-banded vibratory ground motions, resulting from a number of causes including tectonic ground motions, volcanism, landslides, rockbursts, and man-made explosions. Of these, naturally occurring tectonic-related earthquakes are the largest and most important. These are caused by the fracture and sliding of rock along *faults* within the Earth's crust. A fault is a zone of the earth's crust within which the two sides have moved — faults may be hundreds of miles long, from one to over one hundred miles deep, and are sometimes not readily apparent on the ground surface. Earthquakes initiate a number of phenomena or agents, termed *seismic hazards*, which can cause significant damage to the built environment — these include fault rupture, vibratory ground motion (i.e., shaking), inundation (e.g., tsunami, seiche, dam failure), various kinds of permanent ground failure (e.g., liquefaction), fire, or hazardous materials release. In a particular earthquake event, any particular hazard can dominate, and historically each has caused major damage and great loss of life in particular earthquakes.

For most earthquakes, shaking is the dominant and most widespread agent of damage. Shaking near the actual earthquake rupture lasts only during the time when the fault ruptures, a process that takes seconds or at most a few minutes. The seismic waves generated by the rupture propagate long after the movement on the fault has stopped, however, spanning the globe in about 20 min. Typically, earthquake ground motions are powerful enough to cause damage only in the near field (i.e., within a few tens of kilometers from the causative fault) — in a few instances, long period motions have caused significant damage at great distances, to selected lightly damped structures. A prime example of this was the 1985 Mexico City Earthquake, where numerous collapses of mid- and high-rise buildings were due to a magnitude 8.1 Earthquake occurring at a distance of approximately 400 km from Mexico City.

1.2   Causes of Earthquakes and Faulting

In a global sense, tectonic earthquakes result from motion between a number of large plates comprising the earth's crust or lithosphere (about 15 large plates, in total), Figure 1.1.

These plates are driven by the convective motion of the material in the earth's mantle, which in turn is driven by the heat generated at the earth's core. Relative plate motion at the fault interface is constrained by friction and/or *asperities* (areas of interlocking due to protrusions in the fault surfaces). However, strain energy accumulates in the plates, eventually overcomes any resistance, and causes slip between the two sides of the fault. This sudden slip, termed *elastic rebound* by Reid (1910) based on his studies of regional deformation following the 1906 San Francisco Earthquake, releases large amounts of energy, which constitutes or *is* the earthquake. The location of initial radiation of seismic waves (i.e., the first location of dynamic rupture) is termed the *hypocenter*, while the projection on the surface of the earth directly above the hypocenter is termed the *epicenter*. Other terminology includes *near-field*[1] (within one source dimension of the epicenter, where source dimension refers to the width or length of faulting, whichever is shorter), *far-field* (beyond near-field), and *meizoseismal* (the area of strong shaking and damage). Energy is radiated over a broad spectrum of frequencies through the earth, in *body waves* and *surface waves* (Bolt 1993). Body waves are of two types: P waves (transmitting energy via push–pull motion) and slower S waves (transmitting energy via shear action at right angles to the direction of motion). Surface waves are also of two types: horizontally oscillating *Love waves* (analogous to S body waves) and vertically oscillating *Rayleigh waves*.

While the accumulation of strain energy within the plate can cause motion (and consequent release of energy) at faults at any location, earthquakes occur with greatest frequency at the boundaries of the tectonic plates. The boundary of the Pacific plate is the source of nearly half of the world's great earthquakes. Stretching 40,000 km (24,000 miles) around the circumference of the Pacific Ocean, it includes Japan, the west coast of North America, and other highly populated areas, and is aptly termed the *Ring of Fire*. The interiors of plates, such as ocean basins and continental shields, are areas of low seismicity but are not inactive — the largest earthquakes known to have occurred in North America, for example, occurred in 1811–1812 in the New Madrid area, far from a plate boundary. Tectonic plates move relatively slowly (5 cm per year is relatively fast) and irregularly, with relatively frequent small and only occasional large earthquakes. Forces may build up for decades or centuries at plate interfaces until a large movement occurs all at once. These sudden, violent motions produce the shaking that is felt as an earthquake. The shaking can cause direct damage to buildings, roads, bridges, and other man-made structures as well as triggering landslides, fires, tidal waves (tsunamis), and other damaging phenomena.

Faults are the physical expression of the boundaries between adjacent tectonic plates and thus may be hundreds of miles long. In addition, there may be thousands of shorter faults parallel to or branching out from a main fault zone. Generally, the longer a fault the larger the earthquake it can generate. Beyond the main tectonic plates, there are many smaller subplates, "platelets," and simple blocks of crust that occasionally move and shift due to the "jostling" of their neighbors and the major plates. The existence of these many subplates means that smaller but still damaging earthquakes are possible almost anywhere, although often with less likelihood.

---

[1]Not to be confused with *near-source* as used in the 1997 Uniform Building Code, which can be as much as 15 km, depending on type of faulting.

(a)

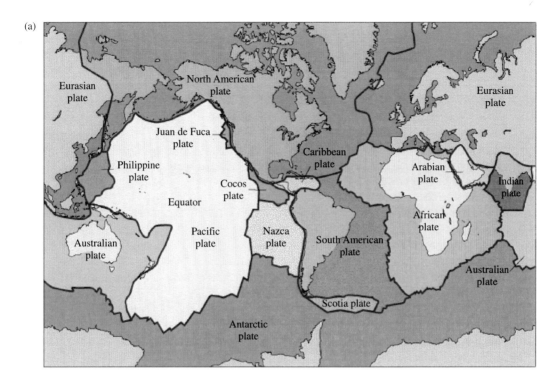

(b)

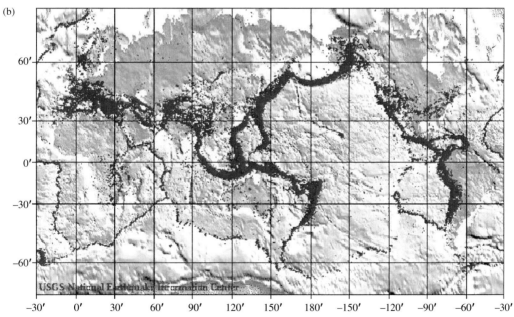

FIGURE 1.1 (a) Global tectonic plate boundaries. (b) Global seismicity 1975–1995 (*from:* U.S. Geological Survey [USGS]).

Faults are typically classified according to their sense of motion, Figure 1.2. Basic terms include *transform* or *strike slip* (relative fault motion occurs in the horizontal plane, parallel to the *strike* of the fault), *dip-slip* (motion at right angles to the strike, up- or down-slip), *normal* (dip-slip motion, two sides in tension, move away from each other), *reverse* (dip-slip, two sides in compression, move toward each other), and *thrust* (low-angle reverse faulting).

Generally, earthquakes will be concentrated in the vicinity of faults, faults that are moving more rapidly than others will tend to have higher rates of seismicity, and larger faults are more likely than others to produce a large event. Many faults are identified on regional geological maps, and useful information on fault location and displacement history is available from local and national geological surveys in areas of high seismicity. Considering this information, areas of an expected large earthquake in the near future (usually measured in years or decades) can, and have, been identified. However, earthquakes continue to occur on "unknown" or "inactive" faults. An important development has been the growing recognition of *blind thrust faults*, which emerged as a result of the several earthquakes in the 1980s, none of which were accompanied by surface faulting (Stein and Yeats 1989). Blind thrust faults are faults at depth occurring under anticlinal folds — since they have only subtle surface expression, their seismogenic potential can only be evaluated by indirect means (Greenwood 1995). Blind thrust faults are particularly worrisome because they are hidden, are associated with folded topography in general, including areas of lower and infrequent seismicity, and, therefore, result in a situation where the potential for an earthquake exists in any area of anticlinal geology, even if there are few or no earthquakes in the historic record. Recent major earthquakes of this type have included the 1980 M_W 7.3 El Asnam (Algeria), 1988 M_W 6.8 Spitak (Armenia), and 1994 M_W 6.7 Northridge (California) events.

Focal mechanism refers to the direction of slip in an earthquake and the orientation of the fault on which it occurs. Focal mechanisms are determined from seismograms and typically displayed on maps as a black and white "beach ball" symbol. This symbol is the projection on a horizontal plane of the lower half of an imaginary, spherical shell (focal sphere) surrounding the earthquake source (USGS, n.d.). A line is scribed where the fault plane intersects the shell. The beach ball depicts the stress-field orientation at the time of rupture such that the black quadrants contain the tension axis (T), which reflects the minimum compressive stress direction, and the white quadrants contain the pressure axis (P), which reflects the maximum compressive stress direction. For mechanisms calculated from first-motion directions (as well as some other methods), more than one focal mechanism solution may fit the data equally well, so that there is an ambiguity in identifying the fault plane on which the slip

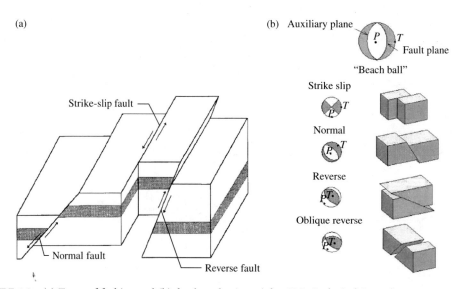

FIGURE 1.2 (a) Types of faulting and (b) focal mechanisms (after U.S. Geological Survey).

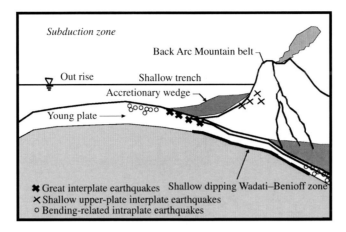

FIGURE 1.3 Schematic diagram of subduction zone, typical of west coast of South America, Pacific Northwest of United States or Japan.

occurred, from the orthogonal, mathematically equivalent, auxiliary plane. The ambiguity may sometimes be resolved by comparing the two fault-plane orientations to the alignment of small earthquakes and aftershocks. The first three examples describe fault motion that is purely horizontal (strike slip) or vertical (normal or reverse). The oblique-reverse mechanism illustrates that slip may also have components of horizontal and vertical motion.

Subduction refers to the plunging of one plate (e.g., the Pacific) beneath another, into the mantle, due to convergent motion, as shown in Figure 1.3. Subduction zones are typically characterized by volcanism, as a portion of the plate (melting in the lower mantle) re-emerges as volcanic lava. Four types of earthquakes are associated with subduction zones: (1) shallow crustal events, in the accretionary wedge; (2) intraplate events, due to plate bending; (3) large interplate events, associated with slippage of one plate past the other; and (4) deep Benioff zone events. Subduction occurs along the west coast of South America at the boundary of the Nazca and South American plate, in Central America (boundary of the Cocos and Caribbean plates), in Taiwan and Japan (boundary of the Philippine and Eurasian plates), and in the North American Pacific Northwest (boundary of the Juan de Fuca and North American plates), among other places.

Probabilistic methods can be usefully employed to quantify the likelihood of an earthquake's occurrence. However, the earthquake generating process is not understood well enough to reliably predict the times, sizes, and locations of earthquakes with precision. In general, therefore, communities must be prepared for an earthquake to occur at any time.

1.3 Measurement of Earthquakes

Earthquakes are complex multidimensional phenomena, the scientific analysis of which requires measurement. Prior to the invention of modern scientific instruments, earthquakes were qualitatively measured by their effect or *intensity*, which differed from point to point. With the deployment of seismometers, an instrumental quantification of the entire earthquake event — the unique *magnitude* of the event — became possible. These are still the two most widely used measures of an earthquake, and a number of different scales for each have been developed, which are sometimes confused.[2]

---

[2]Earthquake magnitude and intensity are analogous to a lightbulb and the light it emits. A particular lightbulb has only one energy level, or wattage (e.g., 100 W, analogous to an earthquake's magnitude). Near the lightbulb, the light intensity is very bright (perhaps 100 ft-candles, analogous to MMI IX), while farther away the intensity decreases (e.g., 10 ft-candles, MMI V). A particular earthquake has only one magnitude value, whereas it has many intensity values.

Engineering design, however, requires measurement of earthquake phenomena in units such as force or displacement. This section defines and discusses each of these measures.

1.3.1 Magnitude

An individual earthquake is a unique release of strain energy — quantification of this energy has formed the basis for measuring the earthquake event. Richter (1935) was the first to define earthquake magnitude, as

$$M_{\mathrm{L}} = \log A - \log A_0 \tag{1.1}$$

where M_{L} is the *local magnitude* (which Richter only defined for Southern California), A is the maximum trace amplitude in micrometers recorded on a standard Wood–Anderson short-period torsion seismometer,[3] at a site 100 km from the epicenter, and $\log A_0$ is a standard value as a function of distance for instruments located at distances other than 100 km and less than 600 km. Subsequently, a number of other magnitudes have been defined, the most important of which are surface wave magnitude M_{S}, body wave magnitude m_{b}, and moment magnitude M_{W}. Due to the fact that M_{L} was only locally defined for California (i.e., for events within about 600 km of the observing stations), surface wave magnitude M_{S} was defined analogously to M_{L}, using teleseismic observations of surface waves of 20 s period (Richter 1935). Magnitude, which is defined on the basis of the amplitude of ground displacements, can be related to the total energy in the expanding wave front generated by an earthquake, and thus to the total energy release — an empirical relation by Richter is

$$\log_{10} E_{\mathrm{S}} = 11.8 + 1.5 M_{\mathrm{S}} \tag{1.2}$$

where E_{S} is the total energy in ergs.[4] Note that $10^{1.5} = 31.6$, so that an increase of one magnitude unit is equivalent to 31.6 times more energy release, two magnitude units increase equivalent to $998.6 \cong 1000$ times more energy, etc. Subsequently, due to the observation that deep-focus earthquakes commonly do not register measurable surface waves with periods near 20 s, a body wave magnitude m_{b} was defined (Gutenberg and Richter 1954), which can be related to M_{S} (Darragh et al. 1994):

$$m_{\mathrm{b}} = 2.5 + 0.63 M_{\mathrm{S}} \tag{1.3}$$

Body wave magnitudes are more commonly used in eastern North America, due to the deeper earthquakes there. A number of other magnitude scales have been developed, most of which tend to *saturate* — that is, asymptote to an upper bound due to larger earthquakes radiating significant amounts of energy at periods longer than used for determining the magnitude (e.g., for M_{S}, defined by measuring 20 s surface waves, saturation occurs at about $M_{\mathrm{S}} > 7.5$). More recently, *seismic moment* has been employed to define a *moment magnitude* M_{W} (Hanks and Kanamori 1979; also denoted as boldface **M**), which is finding increased and widespread use

$$\log M_0 = 1.5 M_{\mathrm{W}} + 16.0 \tag{1.4}$$

where seismic moment M_0 (dyne cm) is defined as (Lomnitz 1974)

$$M_0 = \mu A \bar{u} \tag{1.5}$$

where μ is the material shear modulus, A is the area of fault plane rupture, and $\bar{u}$ is the mean relative displacement between the two sides of the fault (the averaged fault slip). Comparatively, M_{W} and M_{S} are

---

[3]The instrument has a natural period of 0.8 s, critical damping ration 0.8, magnification 2800.

[4]Richter (1958) gives 11.4 for the constant term, rather than 11.8, which is based on subsequent work — the uncertainty in the data makes this difference inconsequential.

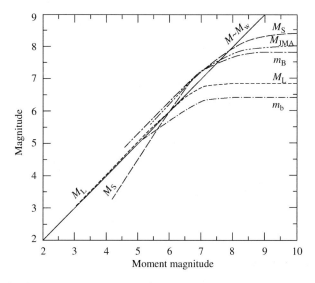

FIGURE 1.4 Relationship between moment magnitude and various magnitude scales (Campbell, K.W. 1985).

numerically almost identical up to magnitude 7.5. Figure 1.4 indicates the relationship between moment magnitude and various magnitude scales.

For lay communications, it is sometimes customary to speak of great earthquakes, large earthquakes, etc. There is no standard definition for these, but the following is an approximate categorization:

Earthquake	Micro	Small	Moderate	Large	Great
Magnitude[a]	Not felt	<5	5–6.5	6.5–8	>8

[a] Not specifically defined.

From the foregoing discussion, it can be seen that magnitude and energy are related to fault rupture length and slip. Slemmons (1977) and Bonilla et al. (1984) have determined statistical relations between these parameters for worldwide and regional data sets, aggregated and segregated by type of faulting (normal, reverse, strike-slip). Bonilla et al.'s worldwide results for all types of faults are

$$M_S = 6.04 + 0.708 \log_{10} L, \qquad s = 0.306 \tag{1.6}$$

$$\log_{10} L = -2.77 + 0.619 M_S, \qquad s = 0.286 \tag{1.7}$$

$$M_S = 6.95 - 0.723 \log_{10} d, \qquad s = 0.323 \tag{1.8}$$

$$\log_{10} d = -3.58 + 0.550 M_S, \qquad s = 0.282 \tag{1.9}$$

which indicates, for example, that, for $M_S = 7$, the average fault rupture length is about 36 km (and the average displacement is about 1.86 m), and s indicates standard deviation. Conversely, a fault of 100 km length is capable of about an $M_S = 7.5$ event[5]. More recently, Wells and Coppersmith (1994) have performed an extensive analysis of a dataset of 421 earthquakes — their results are presented in Table 1.1.

---

[5]Note that $L = g(M_S)$ should not be inverted to solve for $M_S = f(L)$, as a regression for $y = f(x)$ is different than a regression for $x = g(y)$.

TABLE 1.1 Regressions of (a) Rupture Length, Rupture Width, Rupture Area, and Moment Magnitude and (b) Displacement and Moment Magnitude

Equation[a]	Slip type[b]	Number of events	Coefficients and standard errors a(sa)	b(sb)	Standard deviation s	Correlation coefficient r	Magnitude range	Length/width range (km)
			(a) Regressions of rupture length, rupture width, rupture area, and moment magnitude					
M = a + b = log(SRL)	SS	43	5.16(0.13)	1.12(0.08)	0.28	0.91	5.6 to 8.1	1.3–432
	R	19	5.00(0.22)	1.22(0.16)	0.28	0.88	5.4–7.4	3.3–85
	N	13	4.86(0.34)	1.32(0.26)	0.34	0.81	5.2–7.3	2.5–41
	All	77	5.08(0.10)	1.16(0.07)	0.28	0.89	5.2–8.1	1.3–432
log(SRL) = a + b = M	SS	43	-3.55(0.37)	0.74(0.05)	0.23	0.91	5.6–8.1	1.3–432
	R	19	-2.86(0.55)	0.63(0.08)	0.20	0.88	5.4–7.4	3.3–85
	N	15	-2.01(0.65)	0.50(0.10)	0.21	0.81	5.2–7.3	2.5–41
	All	77	-3.22(0.27)	0.69(0.04)	0.22	0.89	5.2–8.1	1.3–432
M = a + b × log(RLD)	SS	93	4.33(0.06)	1.49(0.05)	0.24	0.96	4.8–8.1	1.5–350
	R	50	4.49(0.11)	1.49(0.09)	0.26	0.93	4.8–7.6	1.1–80
	N	24	4.34(0.23)	1.54(0.18)	0.31	0.88	5.2–7.3	3.8–63
	All	167	4.38(0.06)	1.49(0.04)	0.26	0.94	4.8–8.1	1.1–350
log(RLD) = a + b = M	SS	93	-2.57(0.12)	0.62(0.02)	0.15	0.96	4.8–8.1	1.5–350
	R	50	-2.42(0.21)	0.58(0.03)	0.16	0.93	4.8–7.6	1.1–80
	N	24	-1.88(0.37)	0.50(0.06)	0.17	0.88	5.2–7.3	3.8–63
	All	167	-2.44(0.11)	0.59(0.02)	0.16	0.94	4.8–8.1	1.1–350
M = a + b × log(RW)	SS	87	3.80(0.17)	2.59(0.18)	0.45	0.84	4.8–8.1	1.1–350
	R	43	4.37(0.16)	1.95(0.15)	0.32	0.90	4.8–8.1	1.1–80
	N	23	4.04(0.29)	2.11(0.28)	0.31	0.86	5.2–7.3	3.8–63
	All	153	4.06(0.11)	2.25(0.12)	0.41	0.84	4.8–8.1	1.5–350
log(RW) = a + b = M	SS	87	-0.76(0.12)	0.27(0.02)	0.14	0.84	4.8–8.1	1.5–350
	R	43	-1.61(0.20)	0.41(0.03)	0.15	0.90	4.8–7.6	1.1–80
	N	23	-1.14(0.28)	0.35(0.05)	0.12	0.86	5.2–7.3	3.8–63
	All	153	-1.01(0.10)	0.32(0.02)	0.15	0.84	4.8–8.1	1.1–350

Equation	Slip type	Number	a(s.e.)	b(s.e.)	s	r	Magnitude range	Length/Displacement range
M = a + b × log(RA)	SS	83	3.98(0.07)	1.02(0.03)	0.23	0.96	4.8–7.9	3–5184
	R	43	4.33(0.12)	0.90(0.05)	0.25	0.94	4.8–7.6	2.2–2400
	N	22	3.93(0.23)	1.02(0.10)	0.25	0.92	5.2–7.3	19–900
	All	148	4.07(0.06)	0.98(0.03)	0.24	0.95	4.8–7.9	2.2–5184
log(RA) = a + b = M	SS	83	−3.42(0.18)	0.90(0.03)	0.22	0.96	4.8–7.9	3–5184
	R	43	−3.99(0.36)	0.98(0.06)	0.26	0.94	4.8–7.6	2.2–2100
	N	22	−2.87(0.50)	0.82(0.08)	0.22	0.92	5.2–7.3	19–900

(b) Regressions of displacement and moment magnitude

Equation	Slip type	Number	a(s.e.)	b(s.e.)	s	r	Magnitude range	Displacement range
M = a + b × log(MD)	SS	43	6.81(0.06)	0.78(0.06)	0.29	0.90	5.6–8.1	0.01–14.6
	{R[c]	*21*	*6.52(0.11)*	*0.44(0.26)*	*0.52*	*0.36*	*5.4–7.4*	*0.11–6.51}*
	N	16	6.61(0.09)	0.71(0.15)	0.34	0.80	5.2–7.3	0.06–6.1
	All	80	6.69(0.04)	0.74(0.07)	0.40	0.78	5.2–8.1	0.01–14.6
log(MD) = a + b × M	SS	43	−7.03(0.55)	1.03(0.08)	0.34	0.90	5.6–8.1	0.01–14.6
	{R	*21*	*−1.84(1.14)*	*0.29(0.17)*	*0.42*	*0.36*	*5.4–7.4*	*0.11–6.51}*
	N	16	−5.90(1.18)	0.89(0.18)	0.38	0.80	5.2–7.3	0.06–6.1
	All	80	−5.46(0.51)	0.82(0.08)	0.42	0.78	5.2–8.1	0.0–14.6
M = a + b × log(AD)	SS	29	7.04(0.05)	0.89(0.09)	0.28	0.89	5.6–8.1	0.05–8.0
	{R	*15*	*6.64(0.16)*	*0.13(0.36)*	*0.50*	*0.10*	*5.8–7.4*	*0.06–1.51}*
	N	12	6.78(0.12)	0.65(0.25)	0.33	0.64	6.0–7.3	0.08–2.1
	All	56	6.93(0.05)	0.82(0.10)	0.39	0.75	5.6–8.1	0.05–8.0
log(AD) = a + b × M	SS	29	−6.32(0.61)	0.90(0.09)	0.28	0.89	5.6–8.1	0.05–8.0
	{R	*15*	*−0.74(1.40)*	*0.08(0.21)*	*0.38*	*0.10*	*5.8–7.4*	*0.06–1.51}*
	N	12	−4.45(1.59)	0.63(0.24)	0.33	0.64	6.0–7.3	0.08–2.1
	All	56	−4.80(0.57)	0.69(0.08)	0.36	0.75	5.6–8.1	0.05–8.0

[a] SRL—surface rupture length (km); RLD — subsurface rupture length (km); RW — downdip rupture width (km); RA — rupture area (km^2); MD — maximum displacement (m); AD — average displacement (m).

[b] SS— strike slip; R — reverse; N — normal.

[c] Regressions for reverse-slip relationships shown in italics and brackets are not significant at a 95% probability level.

Source: From Wells, D.L. and Coopersmith, K.J. (1994). Empirical Relationships Among Magnitude, Rupture Length, Rupture Width, Rupture Area and Surface Displacements, *Bull. Scismol. Soc. Am,* 84 (4), 974–1002. With permission.

1.3.2 Intensity

In general, seismic intensity is a metric of the effect, or the strength, of an earthquake hazard at a specific location. While the term can be generically applied to engineering measures such as peak ground acceleration (PGA), it is usually reserved for qualitative measures of location-specific earthquake effects, based on observed human behavior and structural damage. Numerous intensity scales developed in preinstrumental times — the most common in use today are the modified Mercalli (MMI) (Wood and Neumann 1931), Rossi–Forel (R–F), Medvedev–Sponheur–Karnik (MSK-64 1981; Grunthal 1998) and its successor the European Macroseismic Scale (EMS-98 1998), and Japan Meteorological Agency (JMA) (Kanai 1983) scales.

Modified Mercalli Intensity (MMI) is a subjective scale defining the level of shaking at specific sites on a scale of I to XII. (MMI is expressed in Roman numerals, to connote its approximate nature.) For example, moderate shaking that causes few instances of fallen plaster or cracks in chimneys constitutes MMI VI. It is difficult to find a reliable relationship between magnitude, which is a description of the earthquake's total energy level, and intensity, which is a subjective description of the level of shaking of the earthquake at specific sites, because shaking severity can vary with building type, design and construction practices, soil type, and distance from the event (Table 1.2).

Note that MMI X is the maximum considered physically possible due to "mere" shaking, and that MMI XI and XII are considered due more to permanent ground deformations and other geologic effects than to shaking.

TABLE 1.2 Modified Mercalli Intensity Scale of 1931 (after Wood and Neumann 1931)

I	Not felt except by a very few under especially favorable circumstances
II	Felt only by a few persons at rest, especially on upper floors of buildings. Delicately suspended objects may swing
III	Felt quite noticeably indoors, especially on upper floors of buildings, but many people do not recognize it as an earthquake. Standing motor cars may rock slightly. Vibration like passing track. Duration estimated
IV	During the day felt indoors by many, outdoors by few. At night some awakened. Dishes, windows, and doors disturbed; walls make creaking sound. Sensation like heavy truck striking building. Standing motorcars rock noticeably
V	Felt by nearly everyone; many awakened. Some dishes, windows, etc., broken; a few instances of cracked plaster; unstable objects overturned. Disturbance of trees, poles, and other tall objects sometimes noticed. Pendulum clocks may stop
VI	Felt by all; many frightened and run outdoors. Some heavy furniture moved; a few instances of fallen plaster or damaged chimneys. Damage slight
VII	Everybody runs outdoors. Damage negligible in buildings of good design and construction slight to moderate in well-built ordinary structures; considerable in poorly built or badly designed structures. Some chimneys broken. Noticed by persons driving motor cars
VIII	Damage slight in specially designed structures; considerable in ordinary substantial buildings, with partial collapse; great in poorly built structures. Panel walls thrown out of frame structures. Fall of chimneys, factory stacks, columns, monuments, walls. Heavy furniture overturned. Sand and mud ejected in small amounts. Changes in well water. Persons driving motor cars disturbed
IX	Damage considerable in specially designed structures; well designed frame structures thrown out of plumb; great in substantial buildings, with partial collapse. Buildings shifted off foundations. Ground cracked conspicuously. Underground pipes broken
X	Some well-built wooden structures destroyed; most masonry and frame structures destroyed with foundations; ground badly cracked. Rails bent. Landslides considerable from river banks and steep slopes. Shifted sand and mud. Water splashed over banks
XI	Few, if any (masonry), structures remain standing. Bridges destroyed. Broad fissures in ground. Underground pipelines completely out of service. Earth slumps and land slips in soft ground. Rails bent greatly
XII	Damage total. Waves seen on ground surfaces. Lines of sight and level distorted. Objects thrown upward into the air

TABLE 1.3   Comparison of Modified Mercalli (MMI) and Other Intensity Scales

a, gals	MMI, Modified Mercalli	R–F, Rossi–Forel	MSK, Medvedev–Sponheur–Karnik	JMA, Japan Meteorological Agency
0.7	I	I	I	0
1.5	II	I–II	II	I
3	III	III	III	II
7	IV	IV–V	IV	II–III
15	V	V–VI	V	III
32	VI	VI–VII	VI	IV
68	VII	VIII–	VII	IV–V
147	VIII	VIII+ to IX–	VIII	V
316	IX	IX+	IX	V–VI
681	X	X	X	VI
(1468)*	XI	—	XI	VII
(3162)*	XII	—	XII	

*a values provided for reference only. MMI > X are due more to geologic effects.

Other intensity scales are defined analogously, Table 1.3, which also contains an approximate conversion from MMI to acceleration a (PGA, in cm/s^2 or gals). The conversion is due to Richter (1935) (other conversions are also available: Trifunac and Brady 1975; Murphy and O'Brien 1977)

$$\log a = \text{MMI}/3 - 1/2 \tag{1.10}$$

Intensity maps are produced as a result of detailed investigation of the type of effects tabulated in Table 1.2, as shown in Figure 1.5 for the 1994 M_W 6.7 Northridge Earthquake. Correlations have been developed between the area of various MMI intensities and earthquake magnitude, which are of value for seismological and planning purposes).

Figure 1.6, for example, correlates A_{felt} versus M_W. For preinstrumental historical earthquakes, A_{felt} can be estimated from newspapers and other reports, which then can be used to estimate the event magnitude, thus supplementing the seismicity catalog. This technique has been especially useful in regions with a long historical record (Ambrayses and Melville 1982; Woo and Muirwood 1984).

1.3.3 Time History

Sensitive strong motion seismometers have been available since the 1930s, and record actual ground motions specific to their location, Figure 1.7. Typically, the ground motion records, termed *seismographs* or *time histories*, have recorded acceleration (these records are termed *accelerograms*) for many years in analog form on photographic film and, more recently, digitally. Analog records required considerable effort for correction due to instrumental drift, before they could be used.

Time histories theoretically contain complete information about the motion at the instrumental location, recording three *traces* or orthogonal records (two horizontal and one vertical). Time histories (i.e., the earthquake motion at the site) can differ dramatically in duration, frequency content, and amplitude. The maximum amplitude of recorded acceleration is termed the peak ground acceleration, PGA (also termed the ZPA, or zero period acceleration) — peak ground velocity (PGV) and peak

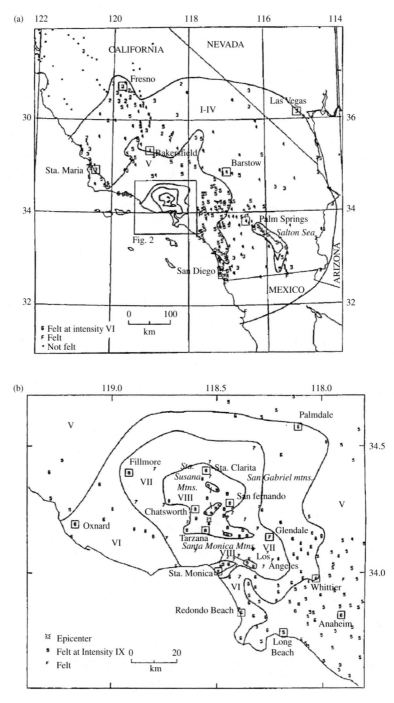

FIGURE 1.5   MMI maps, 1994 M_W 6.7 Northridge Earthquake (courtesy Dewey et al. 1995).

ground displacement (PGD) are the maximum respective amplitudes of velocity and displacement. Acceleration is normally recorded, with velocity and displacement being determined by integration; however, actual velocity and displacement meters are also deployed, to a lesser extent. Acceleration can be expressed in units of cm/s^2 (termed gals), but is often also expressed in terms of the fraction or

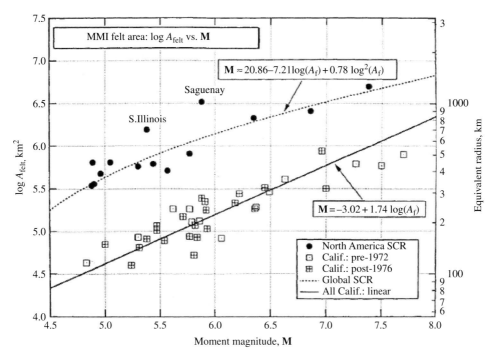

FIGURE 1.6   log A_{felt} (km^2) versus M_W (courtesy Hanks, T.C. and Kanamori, H. 1992).

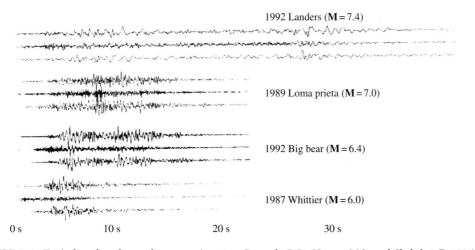

FIGURE 1.7   Typical earthquake accelerograms (courtesy Darragh, R.B., Huang, M.J., and Shakal, A.F. 1994).

percentage of the acceleration of gravity (980.66 gals, termed 1g). Velocity is expressed in cm/s (termed kine). Recent earthquakes (1994 Northridge, M_W 6.7 and 1995 Hanshin [Kobe] M_W 6.9) have recorded PGAs of about 0.8g and PGVs of about 100 kine — almost 2g was recorded in the 1992 Cape Mendocino Earthquake.[6]

---

[6] While almost 2g was recorded in the Cape Mendocino event, the portion of the record was a very narrow spike and while considered genuine, is not considered to be a significant acceleration for structures.

1.3.4 Elastic Response Spectra

If a single degree-of-freedom (SDOF) mass is subjected to a time history of ground (i.e., base) motion similar to that shown in Figure 1.7, the mass or elastic structural response can be readily calculated as a function of time, generating a structural response time history, as shown in Figure 1.8 for several oscillators with differing natural periods. The response time history can be calculated by direct integration of Equation 1.1 in the time domain, or by solution of the Duhamel integral (Clough and Penzien 1975). However, this is time consuming, and the elastic response is more typically calculated

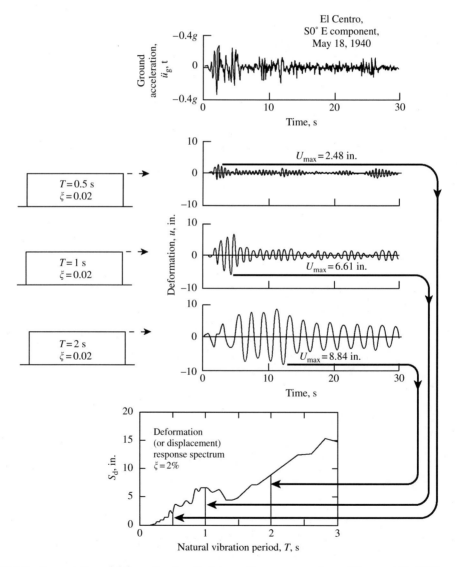

FIGURE 1.8 Computation of deformation (or displacement) response spectrum (Chopra, A.K. 1981).

in the frequency domain

$$\nu(t) = \frac{1}{2\pi} \int_{\varpi=-\infty}^{\infty} H(\varpi)c(\varpi) \exp(i\varpi t)\, d\varpi \tag{1.11}$$

where $\nu(t)$ is the elastic structural displacement response time history, ϖ is the frequency,

$$H(\varpi) = \frac{1}{-\varpi^2 m + ic + k}$$

is the complex frequency response function, and

$$c(\varpi) = \int_{\varpi=-\infty}^{\infty} p(t) \exp(-i\varpi t)\, dt$$

is the Fourier transform of the input motion (i.e., the Fourier transform of the ground motion time history), which takes advantage of computational efficiency using the fast fourier transform (Clough and Penzien 1975).

For design purposes, it is often sufficient to know only the maximum amplitude of the response time history. If the natural period of the SDOF is varied across a spectrum of engineering interest (typically, for natural periods from 0.03 to 3 or more seconds, or frequencies of 0.3 to 30+ Hz), then the plot of these maximum amplitudes is termed a response spectrum. Figure 1.8 illustrates this process, resulting in S_d, the *displacement response spectrum*, while Figure 1.9 shows (a) the S_d, displacement response spectrum, (b) S_v, the *velocity response spectrum* (also denoted PSV, the pseudospectral velocity, pseudo to emphasize that this spectrum is not exactly the same as the relative velocity response spectrum; Hudson, 1979), and (c) S_a, the acceleration *response spectrum*. Note that

$$S_v = \frac{2\pi}{T} S_d = \varpi S_d \tag{1.12}$$

and

$$S_a = \frac{2\pi}{T} S_v = \varpi S_v = \left(\frac{2\pi}{T}\right)^2 S_d = \varpi^2 S_d \tag{1.13}$$

Response spectra form the basis for much modern earthquake engineering structural analysis and design. They are readily calculated *if* the ground motion is known. For design purposes, however, response spectra must be estimated — this process is discussed in another chapter. Response spectra may be plotted in any of several ways, as shown in Figure 1.9 with arithmetic axes, and in Figure 1.10, where the velocity response spectrum is plotted on tripartite logarithmic axes, which equally enables reading of displacement and acceleration response. Response spectra are most normally presented for 5% of critical damping.

While actual response spectra are irregular in shape, they generally have a concave-down arch or trapezoidal shape, when plotted on tripartite log paper. Newmark observed that response spectra tend to be characterized by three regions: (1) a region of constant acceleration, in the high frequency portion of the spectra; (2) constant displacement, at low frequencies; and (3) constant velocity, at intermediate frequencies, as shown in Figure 1.11. If a *spectrum amplification factor* is defined as the ratio of the spectral parameter to the ground motion parameter (where parameter indicates acceleration, velocity, or displacement), then response spectra can be estimated from the data in Table 1.4, provided estimates of the ground motion parameters are available. An example spectrum using these data is given in Figure 1.11.

A standardized response spectrum is provided in the Uniform Building Code (UBC 1997). The spectrum is a smoothed average of a normalized 5% damped spectrum obtained from actual ground motion records grouped by subsurface soil conditions at the location of the recording instrument, and are applicable for earthquakes characteristic of those that occur in California (SEAOC 1988). This normalized shape may be employed to determine a response spectra, appropriate for the soil conditions. Note that the maximum amplification factor is 2.5, over a period range approximately 0.15 s to 0.4–0.9 s, depending on the soil conditions.

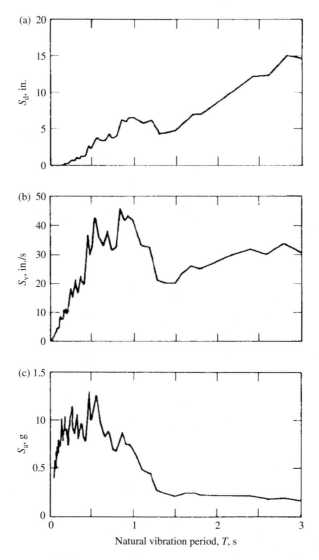

FIGURE 1.9 Response spectra (Chopra, A.K. 1981).

1.3.5 Inelastic Response Spectra

While the foregoing discussion has been for elastic response spectra, most structures are not expected, or even designed, to remain elastic under strong ground motions. Rather, structures are expected to enter the *inelastic* region — the extent to which they behave inelastically can be defined by the ductility factor, μ:

$$\mu = \frac{u_m}{u_y} \tag{1.14}$$

where u_m is the actual displacement of the mass under actual ground motions and u_y is the displacement at yield (i.e., that displacement which defines the extreme of elastic behavior). Inelastic response spectra can be calculated in the time domain by direct integration, analogous to elastic response spectra but with the structural stiffness as a nonlinear function of displacement, $k = k(u)$. If elastoplastic

Response spectrum

Imperial Valley Earthquake
May 18, 1940—2037 PST

III A001 40.001.0 El Centro site
Imperial Valley Irrigation District Comp S0° E
Damping Values are 0, 2, 5, 10, and 20% of critical

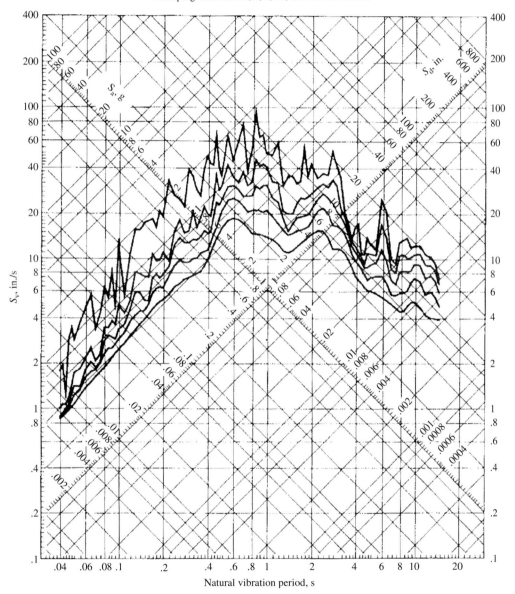

FIGURE 1.10 Response spectra, tripartite plot (El Centro S0° E component) (Chopra, A.K. 1981).

TABLE 1.4   Spectrum Amplification Factors of Horizontal Elastic Response

Damping % critical	One sigma (84.1%)			Median (50%)		
	A	V	D	A	V	D
0.5	5.10	3.84	3.04	3.68	2.59	2.01
1	4.38	3.38	2.73	3.21	2.31	1.82
2	3.66	2.92	2.42	2.74	2.03	1.63
3	3.24	2.64	2.24	2.46	1.86	1.52
5	2.71	2.30	2.01	2.12	1.65	1.39
7	2.36	2.08	1.85	1.89	1.51	1.29
10	1.99	1.84	1.69	1.64	1.37	1.20
20	1.26	1.37	1.38	1.17	1.08	1.01

Source: Newmark, N.M. and Hall, W.J., *Earthquake Spectra and Design,* Earthquake Engineering Research Institute, Oakland, CA, 1982, with permission.

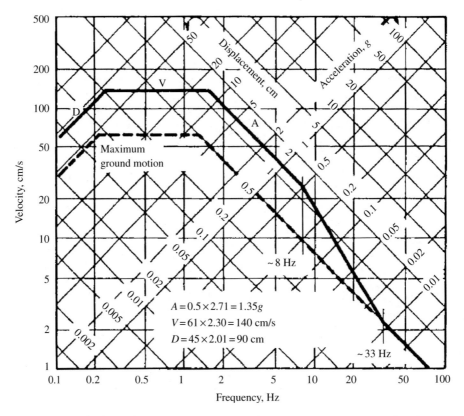

FIGURE 1.11 Idealized elastic design spectrum, horizontal motion (ZPA = 0.5g, 5% damping, one sigma cumulative probability (Newmark, N.M. and Hall, W.J. 1982).

behavior is assumed, then elastic response spectra can be readily modified to reflect inelastic behavior (Newmark and Hall 1982), on the basis that (1) at low frequencies (<0.3 Hz) displacements are the same, (2) at high frequencies (>33 Hz), accelerations are equal, and (3) at intermediate frequencies, the absorbed energy is preserved. Actual construction of inelastic response spectra on this basis is shown in Figure 1.13, where $DVAA_0$ is the elastic spectrum, which is reduced to D' and V' by the ratio of $1/\mu$ for frequencies less than 2 Hz, and by the ratio of $1/(2\mu - 1)^{1/2}$ between 2 and 8 Hz. Above 33 Hz there is no reduction. The result is the inelastic acceleration spectrum ($D'V'A'A_0$), while $A''A_0'$ is the inelastic

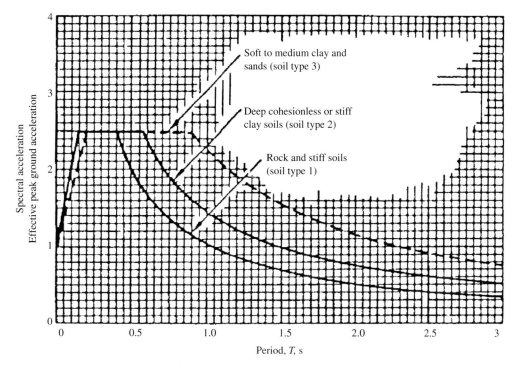

FIGURE 1.12   Normalized response spectra shapes (UBC, 1994).

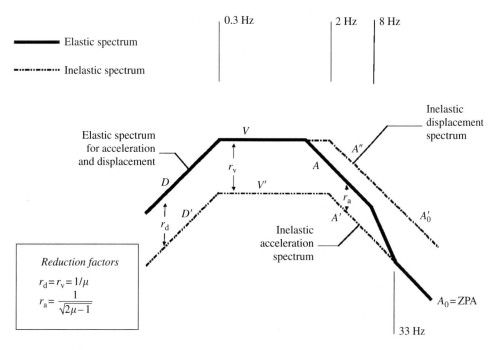

FIGURE 1.13   Inelastic response spectra for earthquakes (Newmark, N.M. and Hall, W.J. 1982).

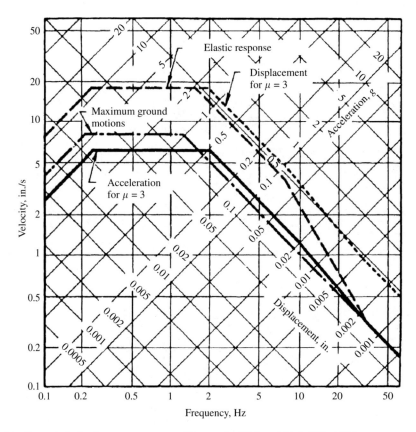

FIGURE 1.14   Example inelastic response spectra (Newmark, N.M. and Hall, W.J. 1982).

displacement spectrum. A specific example, for ZPA $= 0.16g$, damping $= 5\%$ of critical, and $\mu = 3$ is shown in Figure 1.14.

1.4   Distribution of Seismicity

This section discusses and characterizes the nature and distribution of seismicity.

It is evident from Figure 1.1 that some parts of the globe experience more and larger earthquakes than others. The two major regions of seismicity are the circum-Pacific *Ring of Fire* and the *Trans-Alpide belt*, extending from the western Mediterranean through the Middle East and the northern Indian subcontinent to Indonesia. The Pacific plate is created at its South Pacific extensional boundary — its motion is generally northwestward, resulting in relative strike-slip motion in California and New Zealand (with however a compressive component), and major compression and subduction in Alaska, the Aleutians, Kuriles, and northern Japan. Subduction also occurs along the west coast of South America at the boundary of the Nazca and South American plate, in Central America (boundary of the Cocos and Caribbean plates), in Taiwan and Japan (boundary of the Philippine and Eurasian plates), and in the North American Pacific Northwest (boundary of the Juan de Fuca and North American plates). Seismicity in the Trans-Alpide seismic belt is basically due to the relative motions of the African and Australia plates colliding and subducting with the Eurasian plate. The reader is referred to Chen and Scawthorn (2002) for a more extended discussion of global seismicity.

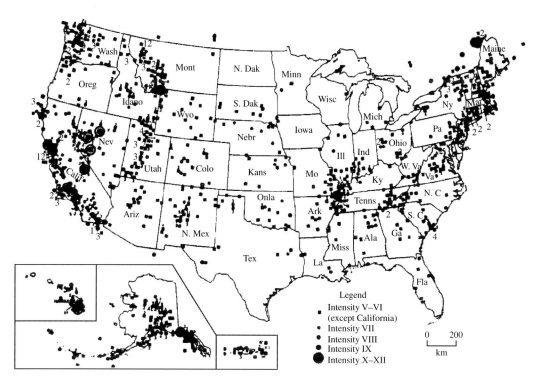

FIGURE 1.15   U.S. seismicity (Algermissen, S.T. 1983; after Coffman et al. 1980).

Regarding U.S. seismicity, the San Andreas fault system in California and the Aleutian Trench off the coast of Alaska are part of the boundary between the North American and Pacific tectonic plates, and are associated with the majority of U.S. seismicity, Figure 1.15. There are many other smaller fault zones throughout the western United States that are also helping to release the stress that is built up as the tectonic plates move past one another, Figure 1.16.

While California has had numerous destructive earthquakes, there is also clear evidence that the potential exists for great earthquakes in the Pacific Northwest (Atwater et al. 1995). On February 28, 2001, the M_W 6.8 Nisqually struck the Puget Sound area, a very similar earthquake to the M_W 6.5 1965 event. Fortunately, the Nisqually event was relatively deep (~50 km), and caused relatively few casualties, although still about $1 billion in damage.

On the east coast of the United States, the cause of earthquakes is less well understood. There is no plate boundary and very few locations of active faults are known so that it is more difficult to assess where earthquakes are most likely to occur. Several significant historical earthquakes have occurred, such as in Charleston, South Carolina, in 1886, and New Madrid, Missouri, in 1811 and 1812, indicating that there is potential for very large and destructive earthquakes (Wheeler et al. 1994; Harlan and Lindbergh 1988). However, most earthquakes in the eastern United States are smaller magnitude events. Because of regional geologic differences, eastern and central U.S. earthquakes are felt at much greater distances than those in the western United States, sometimes up to a thousand miles away (Hopper 1985).

1.5   Strong Motion Attenuation and Duration

The rate at which earthquake ground motion decreases with distance, termed *attenuation*, is a function of the regional geology and inherent characteristics of the earthquake and its source. Three

Earthquake Engineering for Structural Design

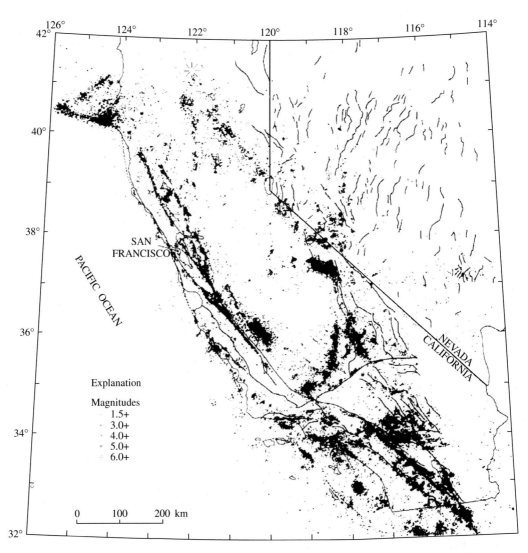

FIGURE 1.16 Seismicity for California and Nevada, 1980–86 **M** > 1.5 (courtesy Jennings, C.W. 1994).

major factors affect the severity of ground shaking at a site: (1) *source* — the size and type of the earthquake; (2) *path* — the distance from the source of the earthquake to the site, and the geologic characteristics of the media earthquake waves pass through; and (3) *site-specific effects* — type of soil at the site. In the simplest of models, if the seismogenic source is regarded as a point then, from considering the relation of energy and earthquake magnitude, and the fact that the volume of a hemisphere is proportion to R^3 (where R represents radius), it can be seen that energy per unit volume is proportional to $C10^{aM} R^{-3}$, where C is a constant or constants dependent on the earth's crustal properties. The constant C will vary regionally — for example, it has long been observed that attenuation in eastern North America (ENA) varies significantly from that in western North America (WNA) — earthquakes in ENA are felt at far greater distances. Therefore, attenuation relations are regionally dependent. Another regional aspect of attenuation is the definition of terms, especially magnitude, where various relations are developed using magnitudes defined by local observatories.

A very important aspect of attenuation is the definition of the distance parameter — since attenuation is the change of ground motion with location, this is clearly important. Many investigators use differing definitions — as study has progressed, several definitions have emerged: (1) hypocentral distance (i.e., straight line distance from point of interest to hypocenter, where hypocentral distance may be arbitrary, or based on regression rather than observation; (2) epicentral distance; (3) closest distance to the causative fault; and (4) closest horizontal distance from the station to the point on the earth's surface that lies directly above the seismogenic source. In using attenuation relations, it is critical that the correct definition of distance is consistently employed.

An extensive discussion of attenuation is beyond the scope of this chapter, and the reader is referred to Chen and Scawthorn for an extended discussion. However, for completeness, we present one attenuation relation, that of Campbell and Bozorgnia (2003 from which the following is excerpted), which can be represented by the expression:

$$\ln Y = c_1 + f_1(M_W) + c_4 \ln \sqrt{f_2(M_W, r_{seis}, S)} + f_3(F) + f_4(S) + f_5(HW, M_W, r_{seis}) + \varepsilon \tag{1.15}$$

where the magnitude scaling characteristics are given by

$$f_1(M_W) = c_2 M_W + c_3(8.5 - M_W)^2 \tag{1.16}$$

The distance scaling characteristics are given by

$$f_2(M_W, r_{seis}, S) = r_{seis}^2 + g(S)^2 \left(\exp\left[c_8 M_W + c_9(8.5 - M_W)^2\right]\right)^2 \tag{1.17}$$

in which the near-source effect of local site conditions is given by

$$g(S) = c_5 + c_6(S_{VFS} + S_{SR})c_7 S_{FR} \tag{1.18}$$

The effect of faulting mechanism is given by

$$f_3(F) = (c_{10} F_{RV} + c_{11} F_{TH}) \tag{1.19}$$

The far-source effect of local site conditions is given by

$$f_4(S) = c_{12} S_{VFS} + c_{13} S_{SR} + c_{14} S_{FR} \tag{1.20}$$

and the effect of the hanging wall (HW) is given by

$$f_5(HW, F, M_W, r_{seis}) = HW f_3(F) f_{HW}(M_W) f_{HW}(r_{seis}) \tag{1.21}$$

where

$$HW = \begin{cases} 0 & \text{for } r_{jb} \geq 5 \text{ km} \quad \text{or} \quad \delta > 70° \\ (S_{VFS} + S_{SR} + S_{FR})(5 - r_{jb})/5 & \text{for } r_{jb} < 5 \text{ km} \quad \text{and} \quad \delta \leq 70° \end{cases} \tag{1.22}$$

$$f_{HW}(M_W) = \begin{cases} 0 & \text{for } M_W < 5.5 \\ M_W - 5.5 & \text{for } 5.5 \leq M_W \leq 6.5 \\ 1 & \text{for } M_W > 6.5 \end{cases} \tag{1.23}$$

and

$$f_{HW}(r_{seis}) = \begin{cases} c_{15}(r_{seis}/8) & \text{for } r_{seis} < 8 \text{ km} \\ c_{15} & \text{for } r_{seis} \geq 8 \text{ km} \end{cases} \tag{1.24}$$

The parameter HW quantifies the effect of the hanging wall and will always evaluate to zero for firm soil and for a horizontal distance of 5 km or greater from the rupture plane. The standard deviation of $\ln Y$ is defined as a function of magnitude according to the expression:

$$\sigma_{\ln Y} = \begin{cases} c_{16} - 0.07 M_W & \text{for } M_W < 7.4 \\ c_{16} - 0.518 & \text{for } M_W \geq 7.4 \end{cases} \tag{1.25}$$

or as a function of PGA according to the expression:

$$\sigma_{\ln Y} = \begin{cases} c_{17} + 0.351 & \text{for } PGA \leq 0.07g \\ c_{17} - 0.132 \ln(PGA) & \text{for } 0.07g < PGA < 0.25g \\ c_{17} + 0.183 & \text{for } PGA \geq 0.25g \end{cases} \tag{1.26}$$

TABLE 1.5 Coefficients for Campbell and Bozorgina Attenuation Relation: Horizontal Component

T_n (s)	c_1	c_2	c_3	c_4	c_5	c_6	c_7	c_8	c_9	c_{10}	c_{11}	c_{12}	c_{13}	c_{14}	c_{15}	c_{16}	c_{17}
Unc PGA	−2.896	0.812	0.000	−1.318	0.187	−0.029	−0.064	0.616	0	0.179	0.307	−0.062	−0.195	−0.320	0.370	0.964	0.263
Cor PGA	−4.033	0.812	0.036	−1.061	0.041	−0.005	−0.018	0.766	0.034	0.343	0.351	−0.123	−0.138	−0.289	0.370	0.920	0.219
0.05	−3.740	0.812	0.036	−1.121	0.058	−0.004	−0.028	0.724	0.032	0.302	0.362	−0.140	−0.158	−0.205	0.370	0.940	0.239
0.075	−3.076	0.812	0.050	−1.252	0.121	−0.005	−0.051	0.648	0.040	0.243	0.333	−0.150	−0.196	−0.208	0.370	0.952	0.251
0.10	−2.661	0.812	0.060	−1.308	0.166	−0.009	−0.068	0.621	0.046	0.224	0.313	−0.146	−0.253	−0.258	0.370	0.958	0.257
0.15	−2.270	0.182	0.041	−1.324	0.212	−0.033	−0.081	0.613	0.031	0.318	0.344	−0.176	−0.267	−0.284	0.370	0.974	0.273
0.20	−2.771	0.812	0.030	−1.153	0.098	−0.014	−0.038	0.704	0.026	0.296	0.342	−0.148	−0.183	−0.359	0.370	0.981	0.280
0.30	−2.999	0.812	0.007	−1.080	0.059	−0.007	−0.022	0.752	0.007	0.359	0.385	−0.162	−0.157	−0.585	0.370	0.984	0.283
0.40	−3.511	0.812	−0.015	−0.964	0.024	−0.002	−0.005	0.842	−0.016	0.379	0.438	−0.078	−0.129	−0.557	0.370	0.987	0.286
0.50	−3.556	0.812	−0.035	−0.964	0.023	−0.002	−0.004	0.842	−0.036	0.406	0.479	−0.122	−0.130	−0.701	0.370	0.990	0.289
0.75	−3.709	0.812	−0.071	−0.964	0.021	−0.002	−0.002	0.842	−0.074	0.347	0.419	−0.108	−0.124	−0.796	0.331	1.021	0.320
1.0	−3.867	0.812	−0.101	−0.964	0.019	0	0	0.842	−0.105	0.329	0.338	−0.073	−0.072	−0.858	0.281	1.021	0.320
1.5	−4.093	0.812	−0.150	−0.964	0.019	0	0	0.842	−0.155	0.217	0.188	−0.079	−0.056	−0.954	0.210	1.021	0.320
2.0	−4.311	0.812	−0.180	−0.964	0.019	0	0	0.842	−0.187	0.060	0.064	−0.124	−0.116	−0.916	0.160	1.021	0.320
3.0	−4.817	0.812	−0.193	−0.964	0.019	0	0	0.842	−0.200	−0.079	0.021	−0.154	−0.117	−0.873	0.089	1.021	0.320
4.0	−5.211	0.812	−0.202	−0.964	0.019	0	0	0.842	−0.209	−0.061	0.057	−0.054	−0.261	−0.889	0.039	1.021	0.320

Note: Uncorrected PGA is to be used only when in estimate of PGA is required. Corrected PGA is to be used when an estimate of PGA compatible with PSA is required.

Source: Adapted from Campbell, K.W. and Bozorgnia, Y. (2003) Updated Near-Source Ground Motion (Attenuation) Relations for the Horizontal and Vertical Components of Peak Ground Acceleration and Acceleration Response Spectra, *Bull. Sesimol. Soc. Am.*

TABLE 1.6   Coefficients for Campbell and Bozorgnia Attenuation Relation: Vertical Component

T_n (s)	c_1	c_2	c_3	c_4	c_5	c_6	c_7	c_8	c_9	c_{10}	c_{11}	c_{12}	c_{13}	c_{14}	c_{15}	c_{16}	c_{17}
Unc PGA	−2.807	0.756	0	−1.391	0.191	0.044	−0.014	0.544	0	0.091	0.223	−0.096	−0.212	−0.199	0.630	1.003	0.320
Cor PGA	−3.108	0.756	0	−1.287	0.142	0.046	−0.040	0.587	0	0.253	0.173	−0.135	−0.138	−0.256	0.630	0.975	0.274
0.05	−1.918	0.756	0	−1.517	0.309	0.069	−0.023	0.498	0	0.058	0.100	−0.195	−0.274	−0.219	0.630	1.031	0.330
0.075	−1.504	0.756	0	−1.551	0.343	0.083	0.000	0.487	0	0.135	0.182	−0.224	−0.303	−0.263	0.630	1.031	0.330
0.10	−1.672	0.756	0	−1.473	0.282	0.062	0.001	0.513	0	0.168	0.210	−0.198	−0.275	−0.252	0.630	1.031	0.330
0.15	−2.323	0.756	0	−1.280	0.171	0.045	0.008	0.591	0	0.223	0.238	−0.170	−0.175	−0.270	0.630	1.031	0.330
0.20	−2.998	0.756	0	−1.131	0.089	0.028	0.004	0.668	0	0.234	0.256	−0.098	−0.041	−0.311	0.571	1.031	0.330
0.30	−3.721	0.756	0.007	−1.028	0.050	0.010	0.004	0.736	0.007	0.249	0.328	−0.026	0.082	−0.265	0.488	1.031	0.330
0.40	−4.536	0.756	−0.015	−0.812	0.012	0	0	0.931	−0.018	0.299	0.317	−0.017	0.022	−0.257	0.428	1.031	0.330
0.50	−4.651	0.756	−0.035	−0.812	0.012	0	0	0.931	−0.043	0.243	0.354	−0.020	0.092	−0.293	0.383	1.031	0.330
0.75	−4.903	0.756	−0.071	−0.812	0.012	0	0	0.931	−0.087	0.295	0.418	0.078	0.091	−0.349	0.299	1.031	0.330
1.0	−4.950	0.756	−0.101	−0.812	0.012	0	0	0.931	−0.124	0.266	0.315	0.043	0.101	−0.481	0.240	1.031	0.330
1.5	−5.073	0.756	−0.150	−0.812	0.012	0	0	0.931	−0.184	0.171	0.211	−0.038	−0.018	−0.518	0.240	1.031	0.330
2.0	−5.292	0.756	−0.180	−0.812	0.012	0	0	0.931	−0.222	0.114	0.115	0.033	−0.022	−0.503	0.240	1.031	0.330
3.0	−5.748	0.756	−0.193	−0.812	0.012	0	0	0.931	−0.238	0.179	0.159	−0.010	−0.047	−0.539	0.240	1.031	0.330
4.0	−6.042	0.756	−0.202	−0.812	0.012	0	0	0.931	−0.248	0.237	0.134	−0.059	−0.267	−0.606	0.240	1.031	0.330

Note: Uncorrected PGA is to be used only when an estimate of PGA is required. Corrected PGA is to be used when an estimate of PGA compatible with PSA is required.

Source: Adapted from Campbell, K.W. and Bozorgnia, Y. (2003) Updated Near-Source Ground Motion (Attenuation) Relations for the Horizontal and Vertical Components of Peak Ground Acceleration and Acceleration Response Spectra, *Bull. Seismol. Soc. Am.*

where PGA is either uncorrected PGA or corrected PGA, depending on the application (see footnote to Table 1.5). The regression coefficients are listed in Table 1.5 and Table 1.6. The relation is considered valid for MW ≥ 4.7 and $r_{seis} \leq 60$ km.

The relation predicts ground motion for firm soil, equivalent to the condition SFS $= 1$, unless one of the site parameters in $g(S)$ and $f_4(S)$ is set to one, in which case it predicts ground motion for either very firm soil, soft rock, or firm rock. The relationship between the faulting mechanism parameters FRV (reverse faulting with dip greater than 45°) and FTH (thrust faulting with dip less than or equal to 45°) and the rake angle λ is[7] (a) strike slip: $F=0$, rake angle $\lambda = 0$–22.5, 177.5–202.5, 337.5–360; (b) normal: $F=0$, rake angle $\lambda = 202.5$–337.5; (c) reverse (FRV $= 1$) $F=1.0$, rake angle $\lambda = 22.5$–157.5 ($\delta > 45$); and (d) thrust (FTH $= 1$) $F=1.0$, rake angle $\lambda = 22.5$–157.5 ($\delta \leq 45$).

Sediment depth D was evaluated and found to be important, but it was not included as a parameter, since it is rarely used in engineering practice. If desired, sediment depth can be included in an estimate of ground motion by using the attenuation relation developed by Campbell (1997, 2000, 2001).

1.6   Characterization of Seismicity

The previous section described the global distribution of seismicity, in qualitative terms. This section describes how that seismicity may be mathematically characterized, in terms of magnitude–frequency and other relations.

The term magnitude–frequency relation was first characterized by Gutenberg and Richter (1954) as

$$\log N(m) = a_N - b_N m \tag{1.27}$$

where $N(m)$ is the number of earthquake events equal to or greater than magnitude m occurring on a seismic source per unit time, and a_N and b_N are regional constants ($10 a_N$ is equal to the total number of earthquakes with magnitude >0, and b_N is the rate of seismicity; b_N is typically 1 ± 0.3). Gutenberg and Richter's examination of the seismicity record for many portions of the earth indicated this relation was valid for selected magnitude ranges. The Gutenberg–Richter relation can be normalized to

$$F(m) = 1 - \exp[-B_M(m - M_0)] \tag{1.28}$$

where $F(m)$ is the cumulative distribution function (CDF) of magnitude, B_M is a regional constant, and M_0 is a small enough magnitude such that lesser events can be ignored. Combining this with a Poisson distribution to model large earthquake occurrence (Esteva 1976) leads to the CDF of earthquake magnitude per unit time:

$$F(m) = \exp[-\exp\{-a_M(m - \mu_M)\}] \tag{1.29}$$

which has the form of a Gumbel (1958) extreme value type I (largest values) distribution (denoted $EX_{I,L}$), which is an unbounded distribution (i.e., the variate can assume any value). The parameters a_M and μ_M can be evaluated by a least squares regression on historical seismicity data, although the probability of very large earthquakes tends to be overestimated. Several attempts have been made to account for this (e.g., Cornell and Merz 1973). Yegulalp and Kuo (1974) have used Gumbel's Type III (largest value, denoted $EX_{III,L}$) to successfully account for this deficiency. This distribution

$$F(m) = \exp\left[-\left(\frac{w - m}{w - u}\right)^k\right] \tag{1.30}$$

has the advantage that w is the largest possible value of the variate (i.e., earthquake magnitude), thus permitting (when w, u, and k are estimated by regression on historical data) an estimate of the source's largest possible magnitude. It can be shown (Yegulalp and Kuo 1974) that estimators of w, u, and k can

---

[7]Rake is a continuous variable representing the angle between the direction of slip on the fault plane and the **strike** or the orientation of the fault on the Earth's surface.

be obtained by satisfying Kuhn–Tucker conditions although, if the data are too incomplete, the $EX_{III,L}$ parameters approach those of the $EX_{I,L}$:

$$u \longrightarrow \mu_M, \quad k/(w - u) \longrightarrow a_M$$

Determination of these parameters requires careful analysis of historical seismicity data (which is highly complex and something of an art; Donovan and Bornstein 1978), and the merging of the resulting statistics with estimates of maximum magnitude and seismicity made on the basis of geological evidence (i.e., as discussed above, maximum magnitude can be estimated from fault length, fault displacement data, time since last event and other evidence, and seismicity can be estimated from fault slippage rates combined with time since last event, see Schwartz, 1988, for an excellent discussion of these aspects). In a full probabilistic seismic hazard analysis, many of these aspects are treated fully or partially probabilistically, including the attenuation, magnitude–frequency relation, upper and lower bound magnitudes for each source zone, geographical bounds of source zones, fault rupture length, and many other aspects. The full treatment requires complex specialized computer codes, which incorporate uncertainty via use of multiple alternative source zonations, attenuation relations, and other parameters (EPRI 1986; Bernreuter et al. 1989) often using a logic tree format. A number of codes have been developed using the public domain FRISK (Fault Risk) code first developed by McGuire (1978).

Several topics are worth noting briefly:

- While analysis of the seismicity of a number of regions indicates that the Gutenberg–Richter relation $\log N(M) = a - bM$ is a good overall model for the magnitude–frequency or probability of occurrence relation, studies of late Quaternary faults during the 1980s indicated that the exponential model is not appropriate for expressing earthquake recurrence on individual faults or fault segments (Schwartz 1988). Rather, it was found that many individual faults tend to generate essentially the same size or *characteristic earthquake* (Schwartz and Coppersmith 1984), having a relatively narrow range of magnitudes at or near the maximum that can be produced by the geometry, mechanical properties, and state of stress of the fault. This implies that, relative to the Gutenberg–Richter magnitude–frequency relation, faults exhibiting characteristic earthquake behavior will have relatively less seismicity (i.e., higher b value) at low and moderate magnitudes, and more near the characteristic earthquake magnitude (i.e., lower b value).

- Most probabilistic seismic hazard analysis models assume the Gutenberg–Richter exponential distribution of earthquake magnitude, and that earthquakes follow a Poisson process, occurring on a seismic source zone randomly in time and space. This implies that the time between earthquake occurrences is exponentially distributed and that the time of occurrence of the next earthquake is independent of the elapsed time since the prior earthquake.[8] The CDF for the exponential distribution is

$$F(t) = 1 - \exp(-\lambda t) \tag{1.31}$$

Note that this forms the basis for many modern building codes, in that the probabilistic seismic hazard analysis results are selected such that the seismic hazard parameter (e.g., PGA) has a "10% probability of exceedance in 50 years" (UBC 1994) — if $t = 50$ years and $F(t) = 0.1$ (i.e., only 10% probability that the event has occurred in t years), then $\lambda = 0.0021$ per year, or 1 per 475 years. A number of more sophisticated models of earthquake occurrence have been investigated, including time-predictable models (Anagnos and Kiremidjian 1984), renewal models (Kameda and Takagi 1981; Nishenko and Buland 1987), and time-dependent models (Ellsworth et al. 1999). The latter have formed the basis for state-of-the-art estimation of seismic hazard for the San Francisco Bay Area by the U.S. Geological Survey, but can only be used when sufficient data are available.

---

[8]For this aspect, the Poisson model is often termed a *memoryless* model.

- Construction of response spectra is usually performed in one of two ways:
  A. Using probabilistic seismic hazard analysis to obtain an estimate of the PGA, and using this to scale a normalized response spectral shape. Alternatively, estimating PGA and PSV (also perhaps PSD) and using these to fit a normalized response spectral shape, for each portion of the spectrum. Since probabilistic response spectra are a composite of the contributions of varying earthquake magnitudes at varying distances, the ground motions of which attenuate differently at different periods, this method has the drawback that the resulting spectra have varying (and unknown) probabilities of exceedance at different periods. Because of this drawback, this method is less favored at present, but still offers the advantage of economy of effort.
  B. An alternative method results in the development of *uniform hazard spectra* (Anderson and Trifunac 1977), and consists of performing the probabilistic seismic hazard analysis for a number of different periods, with attenuation equations appropriate for each period (e.g., those of Boore, Joyner, and Fumal). This method is currently preferred, as the additional effort is not prohibitive, and the resulting response spectra has the attribute that the probability of exceedance is independent of frequency.

The reader is referred to Chen and Scawthorn (2002) for a more extensive discussion of this topic.

Glossary

Attenuation — The rate at which earthquake ground motion decreases with distance.

Benioff zone — A narrow zone, defined by earthquake foci, that is tens of kilometers thick dipping from the surface under the earth's crust to depths of 700 km (also termed Wadat–Benioff zone).

Body waves — Vibrational waves transmitted through the body of the earth, and are of two types: (1) P waves (transmitting energy via dilatational or push-pull motion) and (2) slower S waves (transmitting energy via shear action at right angles to the direction of motion).

Characteristic, earthquake — A relatively narrow range of magnitudes at or near the maximum that can be produced by the geometry, mechanical properties, and state of stress of a fault (Schwartz and Coppersmith 1984).

Completeness — Homogeneity of the seismicity record.

Corner frequency, f_0 — The frequency above which earthquake radiation spectra vary with ϖ^{-3} below f_0, the spectra are proportional to seismic moment.

Cripple wall — A carpenter's term indicating a wood frame wall of less than full height T, usually built without bracing.

Critical damping — The value of damping such that free vibration of a structure will cease after one cycle ($c_{crit} = 2m\omega$). Damping represents the force or energy lost in the process of material deformation (damping coefficient c = force per velocity).

Dip — The angle between a plane, such as a fault, and the earth's surface.

Dip slip — Motion at right angles to the strike, up- or down-slip.

Ductile detailing — Special requirements such as, for reinforced concrete and masonry, close spacing of lateral reinforcement to attain confinement of a concrete core, appropriate relative dimensioning of beams and columns, 135° hooks on lateral reinforcement, hooks on main beam reinforcement within the column, etc.

Ductile frames — Frames required to furnish satisfactory load-carrying performance under large deflections (i.e., ductility). In reinforced concrete and masonry this is achieved by ductile detailing.

Ductility factor — The ratio of the total displacement (elastic plus inelastic) to the elastic (i.e., yield) displacement.

Epicenter — The projection on the surface of the earth directly above the hypocenter.

Far-field — Beyond near-field, also termed teleseismic.

Fault — A zone of the earth's crust within which the two sides have moved — faults may be hundreds of miles long, from one to over one hundred miles deep, and not readily apparent on the ground surface.

Focal mechanism — Refers to the direction of slip in an earthquake, and the orientation of the fault on which it occurs.

Fragility — The probability of having a specific level of damage given a specified level of hazard.

Hypocenter — The location of initial radiation of seismic waves (i.e., the first location of dynamic rupture).

Intensity — A metric of the effect, or the strength, of an earthquake hazard at a specific location, commonly measured on qualitative scales such as MMI, MSK, and JMA.

Lateral force resisting system — A structural system for resisting horizontal forces, due, for example, to earthquake or wind (as opposed to the vertical force resisting system, which provides support against gravity).

Liquefaction — A process resulting in a soil's loss of shear strength, due to a transient excess of pore water pressure.

Magnitude — A unique measure of an individual earthquake's release of strain energy, measured on a variety of scales, of which the moment magnitude M_W (derived from seismic moment) is preferred.

Magnitude–frequency relation — The probability of occurrence of a selected magnitude — the commonest is $\log_{10} n(m) = a - bm$ (Gutenberg and Richter 1954).

Meizoseismal — The area of strong shaking and damage.

Near-field — Within one source dimension of the epicenter, where source dimension refers to the length or width of faulting, whichever is less.

Nonductile frames — Frames lacking ducility or energy absorption capacity due to lack of ductile detailing — ultimate load is sustained over a smaller deflection (relative to ductile frames), and for fewer cycles.

Normal fault — A fault that exhibits dip-slip motion, where the two sides are in tension and move away from each other.

Peak ground acceleration (PGA) — The maximum amplitude of recorded acceleration (also termed the ZPA, or zero period acceleration).

Pounding — The collision of adjacent buildings during an earthquake due to insufficient lateral clearance.

Response spectrum — A plot of maximum amplitudes (acceleration, velocity, or displacement) of a single-degree-of-freedom (SDOF) oscillator as the natural period of the SDOF is varied across a spectrum of engineering interest (typically, for natural periods from 0.03 to 3 or more seconds or frequencies of 0.3 to 30+ Hz).

Reverse fault — A fault that exhibits dip-slip motion, where the two sides are in compression and move away toward each other.

Ring of fire — A zone of major global seismicity due to the interaction (collision and subduction) of the Pacific plate with several other plates.

Sand boils or mud volcanoes — Ejecta of solids (i.e., sand, silt) carried to the surface by water, due to liquefaction.

Seismic gap — A portion of a fault or seismogenic zone that can be deduced to be likely to rupture in the near term, based on patterns of seismicity and geological evidence.

Seismic hazards — The phenomena and/or expectation of an earthquake-related agent of damage, such as fault rupture, vibratory ground motion (i.e., shaking), inundation (e.g., tsunami, seiche, dam failure), various kinds of permanent ground failure (e.g., liquefaction), fire, or hazardous materials release.

Seismic moment — The moment generated by the forces generated on an earthquake fault during slip.

Seismic risk — The product of the hazard and the vulnerability (i.e., the expected damage or loss, or the full probability distribution).

Seismotectonic model — A mathematical model representing the seismicity, attenuation, and related environment.

Soft story — A story of a building signifiantly less stiff than adjacent stories (i.e., the lateral stiffness is 70% or less than that in the story above, or less than 80% of the average stiffness of the three stories above (BSSC 1194).

Spectrum amplification factor — The ratio of a response spectral parameter to the ground motion parameter (where parameter indicates acceleration, velocity, or displacement).

Strike — The intersection of a fault and the surface of the earth, usually measured from north (e.g., the fault strike is N60° W).

Subduction — Refers to the plunging of a tectonic plate (e.g., the Pacific) beneath another (e.g., the North American) down into the mantle, due to convergent motion.

Surface waves — Vibrational waves transmitted within the surficial layer of the earth, and are of two types: horizontally oscillating Love waves (analogous to S body waves) and vertically oscillating Rayleigh waves.

Tectonic — Relating to, causing, or resulting from structural deformation of the earth's crust, (from Greek *tektonikos*, from *tektn*, builder).

Thrust fault — Low-angle reverse faulting (blind thrust faults are faults at depth occurring under anticlinal folds — they have only subtle surface expression).

Trans-alpide belt — A zone of major global seismicity, extending from the Mediterranean through the Middle East, Himalayas, and Indonesian archipelago, resulting from the collision of several major tectonic plates.

Transform or strike-slip fault — A fault where relative fault motion occurs in the horizontal plane, parallel to the strike of the fault.

Uniform hazard spectra — Response spectra with the attribute that the probability of exceedance is independent of frequency.

Vulnerability — The expected damage given a specified value of a hazard parameter.

References

Algermissen, S.T. (1983) *An Introduction to the Seismicity of the United States*, Earthquake Engineering Research Institute, Oakland, CA.

Ambrayses, N.N. and Finkel, C.F. (n.d.) *The Seismicity of Turkey and Adjacent Areas, A Historical Review, 1500–1800*, EREN, Istanbul.

Ambrayses, N.N. and Melville, C.P. (1982) *A History of Persian Earthquakes*, Cambridge University Press, Cambridge.

Anagnos, T. and Kiremidjian, A.S. (1984) Temporal Dependence in Earthquake Occurrence, in *Proc. Eighth World Conf. Earthquake Eng.*, v. 1, Earthquake Engineering Research Institute, Oakland, CA, pp. 255–262.

Anderson, J.G., Trifunac, M.D. (1977) Uniform Risk Absolute Acceleration Spectra, *Advances in Civil Engineering Through Engineering Mechanics: Proceedings of the Second Annual Engineering Mechanics Division Specialty Conference*, Raleigh, NC, May 23–25, 1977; American Society of Civil Engineers, New York, pp. 332–335.

Bernreuter, D.L. et al. (1989) *Seismic Hazard Characterization of 69 Nuclear Power Plant Sites East of the Rocky Mountains*, U.S. Nuclear Regulatory Commission, NUREG/CR-5250.

Bolt, B.A. (1993) *Earthquakes*, W.H. Freeman and Co., New York.

Bonilla, M.G. et al. (1984) Statistical Relations Among Earthquake Magnitude, Surface Rupture Length, And Surface Fault Displacement, *Bull. Seis. Soc. Am.*, 74 (6), 2379–2411.

Campbell, K.W. (1985) Strong Ground Motion Attenuation Relations: A Ten-Year Perspective, *Earthquake Spectra*, 1 (4), 759–804.

Campbell, K.W. (1997) Empirical Near-Source Attenuation Relationships for Horizontal and Vertical Components of Peak Ground Acceleration, Peak Ground Velocity, and Pseudo-Absolute Acceleration Response Spectra, *Seismol. Res. Lett.*, 68, 154–179.

Campbell, K.W. (2000) Erratum: Empirical Near-Source Attenuation Relationships for Horizontal and Vertical Components of Peak Ground Acceleration, Peak Ground Velocity, and Pseudo-Absolute Acceleration Response Spectra, *Seismol. Res. Lett.*, 71, 353–355.

Campbell, K.W. (2001) Erratum: Empirical Near-Source Attenuation Relationships for Horizontal and Vertical Components of Peak Ground Acceleration, Peak Ground Velocity, and Pseudo-Absolute Acceleration Response Spectra, *Seismol. Res. Lett.*, 72, 474.

Campbell, K.W. and Bozorgnia, Y. (2003). Updated Near-Source Ground Motion (Attenuation) Relations for the Horizontal and Vertical Components of Peak Ground Acceleration and Acceleration Response Spectra, *Bull. Seismol. Soc. Am.*, 93 (1), 314–331.

Chen, W.F. and Scawthorn, C. (2002) *Earthquake Engineering Handbook*, CRC Press, Boca Raton.

Chopra, A.K. (1981) *Dynamics of Structures, A Primer,* Earthquake Engineering Research Institute, Oakland, CA.

Clough, R.W. and Penzien, J. (1975) *Dynamics of Structures*, McGraw-Hill, New York.

Coffman, J.L., von Hake, C.A., and Stover, C.W. (1980) *Earthquake History of the United States*, U.S. Dept. of Commerce, NOAA, Pub. 41-1, Washington.

Cornell, C.A. (1968) Engineering Seismic Risk Analysis, *Bull. Seis. Soc. Am.*, 58 (5), 1583–1606.

Cornell, C.A. and Merz, H.A. (1973). Seismic Risk Analysis Based on a Quadratic Magnitude Frequency Law, *Bull. Seis. Soc. Am.*, 63 (6), 1992–2006.

Darragh, R.B., Huang, M.J., and Shakal, A.F. (1994) Earthquake Engineering Aspects of Strong Motion Data from Recent California Earthquakes, *Proc. Fifth U.S. National Conf. Earthquake Engineering*, v. III, Earthquake Engineering Research Institute, Oakland, CA, 99–108.

Dewey, J.W. and Suárez, G. (1991) Seismotectonics of Middle America, in Slemmons, D.B., Engdahl, E.R., Zoback, M.D., and Blackwell, D.B., eds., *Neotectonics of North America*, GSA DNAG Vol., pp. 309–321.

Dewey, J.W. et al. (1995). Spatial Variations of Intensity in the Northridge Earthquake, in Woods, M.C. and Seiple, W.R., eds., *The Northridge California Earthquake of 17 January 1994*, California Department of Conservation, Division of Mines and Geology, Special Publ. 116, Sacramento, pp. 39–46.

Donovan, N.C. and Bornstein, A.E. (1978) Uncertainties in Seismic Risk Procedures, *J. Geotech. Div.*, ASCE 104 (GT7), 869–887.

Earthquake of 17 January (1994) Special Publ. 116, California Department of Conservation, Division of Mines and Geology, Sacramento, pp. 39–46.

Electric Power Research Institute (1986) *Seismic Hazard Methodology for the Central and Eastern United States*, EPRI NP-4726, Menlo Park, CA.

Ellsworth, W.L. et al. (1999) *A Physically-Based Earthquake Recurrence Model for Estimation of Long-Term Earthquake Probabilities*, Workshop on Earthquake Recurrence: State of the Art and Directions for the Future, Istituto Nazionale de Geofisica, Rome, Italy, February.

Esteva, L. (1976) Seismicity, in Lomnitz, C. and Rosenblueth, E., eds., *Seismic Risk and Engineering Decisions*, Elsevier, New York.

European Seismological Commission (1998) European Macroseismic Scale 1998, EMS-98, Grunthal, G. editor, Subcommission on Engineering Seismology, Working Group Macroseismic Scales, Geo ForschungsZentrum Potsdam, Germany, http: www.gfz-potsdam.de/pb1/pg2/ems_new/INDEX.HTM

Greenwood, R.B. (1995) Characterizing blind thrust fault sources — an overview, in Woods, M.C. and W.R. Seiple., eds., *The Northridge California Earthquake of 17 January 1994*, Calif. Dept. Conservation, Div. Mines and Geology, Special Publ. 116, pp. 279–287.

Grunthal G. (1998) European Macroseismic Scale, Cahiers du Centre Europeen de Geodynamique et de Seismologie, pp. 1–99.

GSHAP North Andes (1998) *Global Seismic Hazard Assessment Program, North Andean Region Final Report. Index Final Report*, Observatorio de San Calixto, Bolivia Instituto de Investigaciones en Geociencias, Minería y Química (INGEOMINAS), Colombia Escuela Politécnica de Quito (EPN), Ecuador Instituto Geofísico del Perú (IGP), Fundación Venezolana de Investigaciones

Sismológica (FUNVISIS), Venezuela GeoforschungsZentrum (GFZ), Germany Institute of Geophysics, ETH, Switzerland Istituto Nazionale di Geofisica (ING), Italy; http://seismo.ethz. ch/gshap/piloto/report.html

Gumbel, E.J. (1958) *Statistics of Extremes*, Columbia University Press, New York.

Gutenberg, B. and Richter, C.F. (1954) *Seismicity of the Earth and Associated Phenomena*, Princeton University Press, Princeton.

Hanks, T.C. and Johnston, A.C. (1992) Common Features of the Excitation and Propagation of Strong Ground Motion for North American Earthquakes, *Bull. Seis. Soc. Am.*, 82 (1), 1–23.

Hanks, T.C. and Kanamori, H. (1979) A Moment Magnitude Scale, *J. Geophys. Res.*, 84, 2348–2350.

Harlan, M.R. and Lindbergh, C. (1988) An Earthquake Vulnerability Analysis of the Charleston, South Carolina, Area, Rept. No. CE-88-1, Dept. of Civil Engng, The Citadel, Charleston, SC.

Hopper, M.G. (1985) Estimation of Earthquake Effects associated with Large Earthquakes in the New Madrid Seismic Zone, U.S.G.S. Open File Report 85-457, Washington.

Hudson, D.E. (1979) *Reading and Interpreting Strong Motion Accelerograms*, Earthquake Engineering Research Institute, Oakland, CA.

IAEE (1992) *Earthquake Resistant Regulations: A World List-1992*. Rev. ed. Prepared by the International Association for Earthquake Engineering. Tokyo: International Association for Earthquake Engineering, 1992. Approximately 1100 pages. Distributed by Gakujutsu Bunken Fukyu-Kai (Association for Science Documents Information) Oh-Okayama, 2-12-1, Meguroku, Tokyo, 152, Japan.

Jennings, C.W. (1994) *Fault Activity Map of California and Adjacent Areas*, Dept. of Conservation, Div. Mines and Geology, Sacramento.

Kameda, H. and Takagi, H. (1981) *Seismic Hazard Estimation Based on Non-Poisson Earthquake Occurrences*, Mem. Fac. Engng, Kyoto Univ., v. XLIII, Pt. 3, July, Kyoto.

Kanai, K. (1983) *Engineering Seismology*, University of Tokyo Press, Tokyo.

Lomnitz, C. (1974). *Global Tectonics and Earthquake Risk*, Elsevier, New York.

McGuire, R.K. (1978) *FRISK: Computer Program for Seismic Risk Analysis Using Faults as Earthquake Sources*, US Geological Survey, Reports, United States Geological Survey Open file 78-1007, 71 pp.

McGuire, R.K., ed. (1993) *Practice of Earthquake Hazard Assessment*. International Association of Seismology and Physics of the Earth's Interior, 284 pp.

MSK-64 (1981) Meeting on Up-dating of MSK-64. Report on the Ad-hoc Panel Meeting of Experts on Up-dating of the MSK-64 Seismic Intensity Scale, Jene, 10–14 March 1980, Gerlands Beitr. Geophys., Leipzig 90, 3, 261–268.

Murphy J.R. and O'Brien, L.J. (1977) The Correlation of Peak Ground Acceleration Amplitude with Seismic Intensity and Other Physical Parameters, *Bull. Seis. Soc. Am.*, 67 (3), 877–915.

Newmark, N.M. and Hall, W.J. (1982) *Earthquake Spectra and Design*, Earthquake Engineering Research Institute, Oakland, CA.

Nishenko, S.P. and Buland, R. (1987) A Generic Recurrence Interval Distribution For Earthquake Forecasting, *Bull. Seis. Soc. Am.*, 77, 1382–1399.

Reid, H.F. (1910) The Mechanics of the Earthquake, The California Earthquake of April 18, 1906, Report of the State Investigation Committee, v. 2, Carnegie Institution of Washington, Washington, D.C.

Richter, C.F. (1935) An Instrumental Earthquake Scale, *Bull. Seis. Soc. Am.*, 25, 1–32.

Richter, C.F. (1958) *Elementary Seismology*, W.H. Freeman, San Francisco.

Scholz, C.H. (1990) *The Mechanics of Earthquakes and Faulting*, Cambridge University Press, New York.

Schwartz, D.P. (1988) Geologic Characterization of Seismic Sources: Moving into the 1990s, in J.L. v. Thun, ed., *Earthquake Engineering and Soil Dynamics II — Recent Advances in Ground-Motion Evaluation*, Geotechnical Spec. Publ. No. 20., American Soc. Civil Engrs., New York.

Schwartz, D.P. and Coppersmith, K.J. (1984) Fault Behavior and Characteristic Earthquakes: Examples from the Wasatch and San Andreas Faults, *J. Geophys. Res.*, 89, 5681–5698.

SEAOC (1980) *Recommended Lateral Force Requirements and Commentary*, Seismology Committee, Structural Engineers of California, San Francisco, CA.

Slemmons, D.B. (1977) State-of-the-Art for Assessing Earthquake Hazards in the United States, Report 6: Faults and Earthquake Magnitude, U.S. Army Corps of Engineers, Waterways Experiment Station, Misc. Paper s-73-1, 129 pp.

Stein, R.S. and Yeats, R.S. (1989) Hidden Earthquakes, *Sci. Am.*, June, 260, 48–57.

Structural Engineers Association of California (1988) *Recommended Lateral Force Requirements and Tentative Commentary*, Structural Engineers Association of California, San Francisco, CA.

Trifunac, M.D. and Brady, A.G. (1975) A Study on the Duration of Strong Earthquake Ground Motion, *Bull. Seis. Soc. Am.*, 65, 581–626.

Uniform Building Code (1994). *Volume 2, Structural Engineering Design Provisions*, Intl. Conf. Building Officials, Whittier.

Uniform Building Code (1997) *Volume 2, Structural Engineering Design Provisions*, Intl. Conf. Building Officials, Whittier.

Wells, D.L. and Coppersmith, K.J. (1994) Empirical Relationships among Magnitude, Rupture Length, Rupture Width, Rupture Area and Surface Displacement, *Bull. Seis. Soc. Am.*, 84 (4), 974–1002.

Wheeler, R.L. et al. (1994) Elements of Infrastructure and Seismic Hazard in the Central United States, U.S.G.S. Prof. Paper 1538-M, Washington.

Woo, G., Wood, H.O., and Muir R. (1984) British Seismicity and Seismic Hazard, in *Proceedings of the Eighth World Conference on Earthquake Engineering*, v. I, Earthquake Engineering Research Institute, Oakland, CA, pp. 39–44.

Wood, H.O. and Neumann, Fr. (1931) Modified Mercalli Intensity Scale of 1931, *Bull. Seis. Soc. Am.*, 21, 277–283.

Working Group on California Earthquake Probabilities (1999) Earthquake Probabilities in the San Francisco Bay Region: 2000 to 2030 — A Summary of Findings, Open-File Report 99-517, US Geological Survey, Washington.

Yegulalp, T.M. and Kuo, J.T. (1974) Statistical Prediction of the Occurrence of Maximum Magnitude Earthquakes, *Bull. Seis. Soc. Am.*, 64 (2), 393–414.

Youngs, R.R. and Coppersmith, K.J. (1989) Attenuation Relationships for Evaluation of Seismic Hazards from Large Subduction Zone Earthquakes. *Proceedings of Conference XLVIII: 3rd Annual Workshop on Earthquake Hazards in the Puget Sound, Portland Area*, March 28–30, Portland, Oregon; Hays-Walter-W, Ed. US Geological Survey, Reston, VA, 1989, pp. 42–49.

Youngs, R.R. and Coppersmith, K.J. (1987) Implication of Fault Slip Rates and Earthquake Recurrence Models to Probabilistic Seismic Hazard Estimates, *Bull. Seis. Soc. Am.*, 75, 939–964.

Further Reading

There is a plethora of good references on earthquakes. Chen and Scawthorn (2002) provides an extensive reference. The reader is recommended to *Earthquakes* by B.A. Bolt (1993, Freeman, San Francisco) for an excellent and readable introduction to the subject; to *The Mechanics of Earthquakes and Faulting* by C.A. Scholz (1990, Cambridge University Press, New York) for an erudite treatment of seismogenesis; to *The Geology of Earthquakes* by R.S. Yeats, K. Sieh, and C.R. Allen (1997, Oxford University Press, New York) for an exhaustive review of faulting around the world; and to *Modern Global Seismology* by T. Lay and T.C. Wallace (1995, Academic Press, New York) for a readable theoretical text on seismology (a very rare thing).

2

Earthquake Damage to Structures

Mark Yashinsky
Division of Structures Design,
California Department of
Transportation,
Sacramento, CA

2.1 Introduction

2.1.1 Earthquakes

Most earthquakes occur due to the movement of faults. Faults slowly build up stresses that are suddenly released during an earthquake. We measure the size of earthquakes using moment magnitude as defined in Equation 2.1:

$$M = \left(\frac{2}{3}\right)[\log(M_0) - 16.05] \tag{2.1}$$

where M_0 is the seismic moment as defined in Equation 2.2:

$$M_0 = GAD \quad \text{(in dyne cm)} \tag{2.2}$$

where G is the shear modulus of the rock (dyne/cm^2), A is the area of the fault (cm^2), and D is the amount of slip or movement of the fault (cm).

The largest-magnitude earthquake that can occur on a particular fault is the product of the fault length times its depth (A), the average slip rate times the recurrence interval of the earthquake (D), and the hardness of the rock (G).

For instance the northern half of the Hayward Fault (in the San Francisco Bay Area) has an annual slip rate of 9 mm/year (Figure 2.1). It has an earthquake recurrence interval of 200 years. It is 50 km long and 14 km deep. G is taken as 3×10^{11} dyne/cm^2.

$$M_0 = (0.9 \times 200)(5 \times 10^6)(1.4 \times 10^6)(3 \times 10^{11}) = 3.78 \times 10^{26}$$
$$M = (2/3)[\log 3.78 \times 10^{26} - 16.05] = 7.01$$

Therefore, about a magnitude 7.0 earthquake is the maximum event that can occur on the northern section of the Hayward Fault. Since G is a constant, the average slip is usually a few meters, and

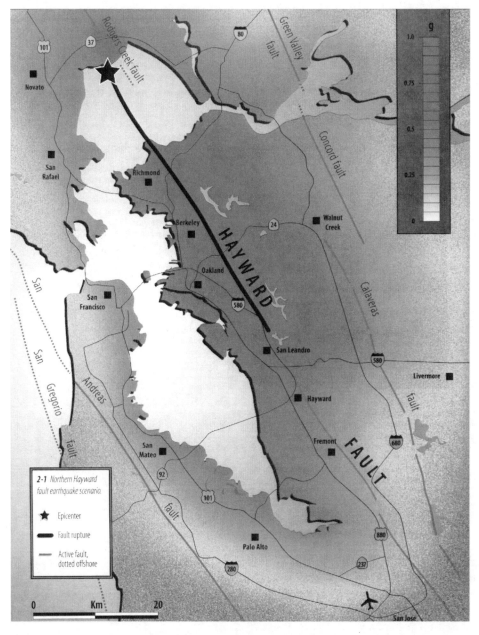

FIGURE 2.1 Map of Hayward Fault (courtesy of EERI, Earthquake Engineering Research Institute, HF-96, The Institute, Oakland, CA, 1996).

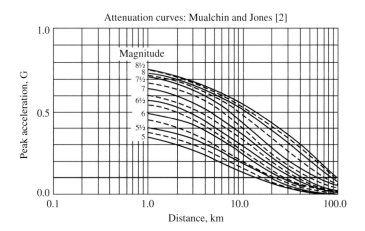

FIGURE 2.2 Attenuation curve developed by Mualchin and Jones [2].

the depth of the crust is fairly constant, the size of the earthquake is usually controlled by the length of the fault.

Magnitude is not particularly revealing to the structural engineer. Engineers design structures for the peak accelerations and displacements at the site. After every earthquake, seismologists assemble the recordings of acceleration versus distance to create attenuation curves that relate the peak ground acceleration (PGA) to the magnitude of earthquakes with distance from the fault rupture (Figure 2.2).

All of the data available on active faults is assembled to create a seismic hazard map. The map has contour lines that provide the peak acceleration based on attenuation curves that provide the reduction in acceleration due to the distance from a fault. The map is based on deterministic derived earthquakes or on earthquakes with the same return period.

2.1.2 Structural Damage

Every day, regions of high seismicity experience many small earthquakes. However, structural damage does not usually occur until the magnitude approaches 5.0. Most structural damage during earthquakes is caused by the failure of the surrounding soil or from strong shaking. Damage also results from surface ruptures, the failure of nearby lifelines, or the collapse of more vulnerable structures. We consider these effects secondary because they are not always present during an earthquake. However, when there is a long surface rupture such as that which accompanied the 1999 Ji Ji, Taiwan, earthquake, secondary effects can dominate.

Since damage can mean anything from minor cracks to total collapse, categories of damage have been developed as shown in Table 2.1. These levels of damage give engineers a choice for the performance of their structure during earthquakes. Most engineered structures are designed only to prevent collapse. This is not only to save money, but also because as a structure becomes stronger it attracts larger forces. Thus, most structures are designed to have sufficient ductility to survive an earthquake. This means that elements will yield and deform but they will be strong in shear and continue to support their load during and after the earthquake.

As shown in Table 2.1, the time that is required to repair damaged structures is an important parameter that weighs heavily on the decision making process. When a structure must be quickly repaired or must remain in service, a different damage state should be chosen.

During large earthquakes the ground is jerked back and forth, causing damage to the element whose capacity is furthest below the earthquake demand. Figure 2.3 shows that the cause may be the supporting soil, the foundation, weak flexural or shear elements, or secondary hazards such as surface

TABLE 2.1   Categories of Structural Damage

Damage state	Functionality	Repairs required	Expected outage
None (preyield) (1)	No loss	None	None
Minor/slight (2)	Slight loss	Inspect, adjust, patch	<3 days
Moderate (3)	Some loss	Repair components	<3 weeks
Major/extensive (4)	Considerable loss	Rebuild components	<3 months
Complete/collapse (5)	Total loss	Rebuild structure	>3 months

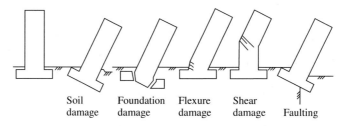

Soil   Foundation   Flexure   Shear
damage   damage   damage   damage   Faulting

FIGURE 2.3   Common types of damage during large earthquakes.

faulting or failure of a nearby structure. Damage also frequently occurs due to the failure of connections, from large torsional moments, from tension and compression, buckling, pounding, etc.

In this chapter structural damage as a result of soil problems, structural shaking, and secondary causes will be discussed. These types of damage illustrate the most common structural hazards that have been seen during recent earthquakes.

2.2   Damage as a Result of Problem Soils

2.2.1   Liquefaction

One of the most common causes of damage to structures is the result of liquefaction to the surrounding soil. When loose saturated sands, silts, or gravel are shaken, the material consolidates, reducing the porosity and increasing pore water pressure. The ground settles, often unevenly, tilting and toppling structures that were formerly supported by the soil. During the 1955 Niigata, Japan, earthquake, several four-story apartment buildings toppled over due to liquefaction (Figure 2.4).

These buildings fell as the liquefied soil lost its ability to support them. As can be seen clearly in Figure 2.5, there was little damage to these buildings and it was reported that their collapse took place over several hours.

Partial liquefaction of the soil in Adapazari during the 1999 Kocaeli, Turkey, earthquake caused several buildings to settle or fall over. Figure 2.6 shows a building that settled as pore water was pushed to the surface, reducing the bearing capacity of the soil. Note that the weight of the building squeezed the weakened soil under the adjacent roadway. Another problem during liquefaction is that the increased pore pressure pushes quay walls, riverbanks, and the piers of bridges toward adjacent bodies of water, often dropping the end spans in the process.

The Shukugawa Bridge is a three-span, continuous, steel box girder superstructure with a concrete deck. The end spans are 87.5 m and the center span is 135 m. The superstructure is supported by steel, multicolumn bents with dropped-bent caps. It is part of a long elevated viaduct, and has expansion joints at Pier 131 and Pier 134. The columns are supported by steel piles embedded in reclaimed land along Osaka Bay.

During the 1995 Kobe, Japan, earthquake, increased pore pressure pushed the quay wall near the west end of the bridge toward the river, allowing the soil and westernmost pier (Pier 134) to move 1 m

FIGURE 2.4   Liquefaction caused building failure in Niigata, Japan. (Photo by Joseph Penzien; photo courtesy of Steinbrugge Collection, Earthquake Engineering Research Center, University of California, Berkeley.)

FIGURE 2.5   Liquefaction caused building failure in Niigata, Japan. (Photo by Joseph Penzien; photo courtesy of Steinbrugge Collection, Earthquake Engineering Research Center, University of California, Berkeley.)

eastward (Figure 2.7). This resulted in the girders falling off their bearings, damaging the expansion joint devices and making the bridge inaccessible. The easternmost pier (Pier 131) moved half a meter toward the river. It appears that the restrainers were the only thing that kept the superstructure at the expansion joint above Pier 134 together, preventing the collapse of the west span. The expansion joint had a 0.6-m vertical offset, and excavation showed that the piles at Pier 134 were also damaged due to the longitudinal movement.

Structures supported on liquefied soil topple, structures that retain liquefied soil are pushed forward, and structures buried in liquefied soil (like culverts and tunnels) float to the surface in the newly buoyant medium.

FIGURE 2.6   Settlement of building due to loss of bearing during the 1999 Kocaeli Earthquake.

FIGURE 2.7   Liquefaction caused bridge damage during the Kobe Earthquake.

The Webster and Posey Street Tube Crossings are 4500-ft-long tubes carrying two lanes of traffic under the Oakland, California, Estuary. The Posey Street Tube was built in the 1920s (Figure 2.8) while the Webster Street Tube was built in the 1960s (Figure 2.9). They are reinforced concrete (RC) tubes with a bituminous coating for waterproofing. The ground was excavated and each tube section was joined to the previously laid section. The tube descends to 70 ft below sea level.

During the 1989 Loma Prieta, California, earthquake, the soil surrounding the Webster and Posey Tubes (that carry traffic through the Oakland Estuary) liquefied. The tunnels began to float to the surface, breaking the joints between sections and slowly filling with water (Figure 2.10 and Figure 2.11).

2.2.2 Landslides

When a steeply inclined mass of soil is suddenly shaken, a slip-plane can form and the material slides downhill. During a landslide, structures sitting on the slide move downward and structures below the slide are hit by falling debris (Figure 2.12).

Landslides frequently occur in canyons, along cliffs, on mountains, and anywhere else where unstable soil exists. Landslides can occur without earthquakes (they often occur during heavy rains that increase the weight and reduce the friction of the soil) but the number of landslides is greatly increased wherever large earthquakes occur. Landslides can move a few inches or hundreds of feet. They can be the result of liquefaction, weak clays, erosion, subsidence, ground shaking, etc.

During the 1999 Ji Ji, Taiwan, earthquake, many of the mountain slopes were denuded by slides, which continued to be a hazard for people traveling on mountain roads in the weeks following the earthquake. The many RC gravity retaining walls that supported the road embankments in the mountainous terrain were all damaged: either from being pushed downhill by the slide (Figure 2.13) or in some cases broken when the retaining wall was restrained from moving downhill (Figure 2.14).

One of the more interesting retaining wall failures during the Ji Ji Earthquake was in a geogrid fabric and mechanically stabilized earth (MSE) wall at the entrance to the Southern International University (Figure 2.15). This wall was quite long and tall and its failure was a surprise since MSE walls have a good performance record during earthquakes. It was speculated that the geogrid retaining system had insufficient embedment into the soil and also it was unclear why an MSE wall would be used in a cut roadway section.

One of the best-known and largest landslides occurred at Turnagain Heights in Anchorage during the 1964 Great Alaska Earthquake. The area of the slide was about 8500 ft wide by 1200 ft long. The average drop was about 35 ft. This slide was complex, but the main cause was the failure of the weak clay layer and the unhindered movement of the ground down the wet mud flats to the sea. Figure 2.16 and Figure 2.17 provide a section and plan view of the slide.

The soil failed due to the intense shaking, and the whole neighborhood of houses, schools, and other buildings slid hundreds of yards downhill, many remaining intact during the fall (Figure 2.18).

Bridges are also severely damaged by landslides. During the 1999 Ji Ji, Taiwan, earthquake, landslides caused the collapse of two bridges. The Tsu Wei Bridges were two parallel three-span structures that crossed a tributary of the Dajia River near the city of Juolan. The superstructure was simply supported "T" girders on hammerhead single-column bents with "drum"-type footings and seat-type abutments. The girders sat on elastomeric pads between transverse shear keys. The spans were about 80 ft long by 46 ft wide, and had a 30° skew. The head scarp was clearly visible on the hillside above the bridge. During the earthquake, the south abutment was pushed forward by the landslide, the first spans fell off the bent caps on the (far) north side, and the second span of the left bridge also fell off of the far bent cap (Figure 2.19).

Also, the tops of the columns at Bent 2 had rotated away from the (south) Abutment 1. Therefore, it appears that both the top of Abutment 1 and the top of Bent 2 had moved away from the slide, while the remaining spans, restrained by Bent 3 and Abutment 4, had remained in place. Perhaps the landslide originally had pushed against Bent 2, rotating the columns forward, and the debris had

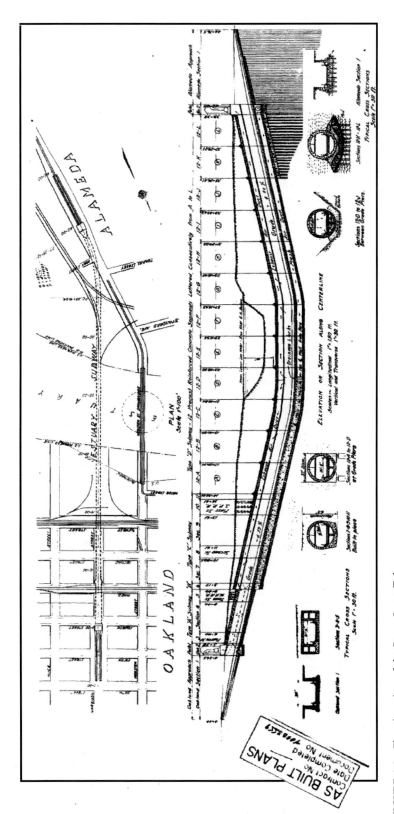

FIGURE 2.8 Elevation view of the Posey Street Tube.

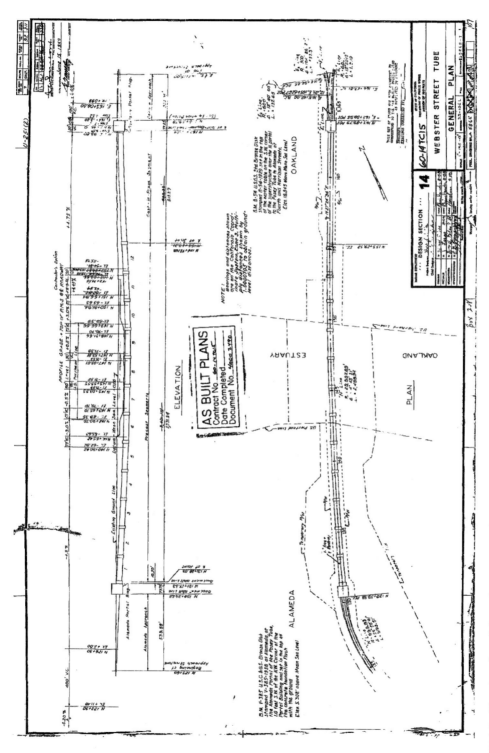

FIGURE 2.9 Elevation view of the Webster Street Tube.

FIGURE 2.10   Liquefaction induced damage to the Webster Street Tube tunnel.

FIGURE 2.11   Liquefaction induced damage to the Webster Street Tube tunnel.

Before landslide

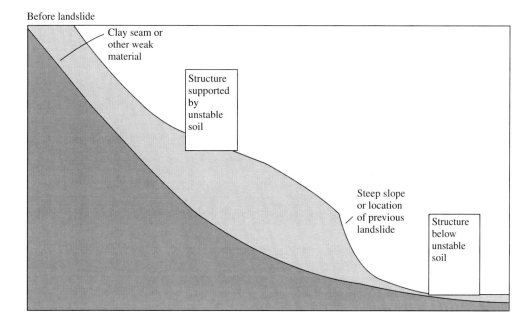

After landslide

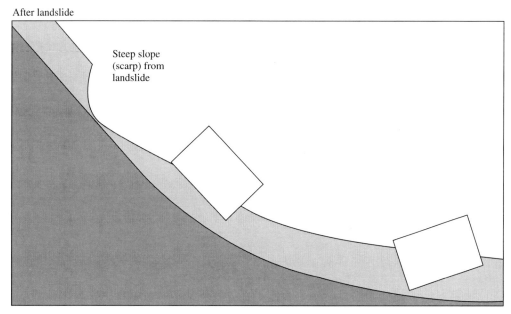

FIGURE 2.12 Diagram showing typical features of landslides.

since been removed by the current or by a construction crew. Perhaps the skew had rotated the spans to the right as they fell, pushing them against the shear keys at Bent 2, which rotated the top of the columns forward and eventually pushed the spans off the tops of Bent 2 and Bent 3. Or perhaps there was an element of strong shaking that combined with the landslide to create the column rotation and fallen spans.

Dams are particularly vulnerable to landslides since they are frequently built to hold back the water in canyons and mountain streams. Moreover, inspection of the dam after an earthquake is often difficult when slides block the roads leading to the dam. When the Pacoima concrete arch dam was built in the

FIGURE 2.13   Gravity retaining wall pushed outward by landslide.

1920s, a covered tunnel was constructed to allow access to the dam. However this tunnel, along with roads and a tramway to the dam, were damaged by massive landslides during and for several days after the 1971 San Fernando Earthquake (Figure 2.20).

The Lower San Fernando Dam for the Van Norman Reservoir was also severely damaged during the 1971 San Fernando Earthquake. It was fortunate that water levels were low as the concrete crest on this earthen dam collapsed due to a large landslide along both the upstream (Figure 2.21) and the downstream (Figure 2.22) faces. Considering the vulnerability of thousands of residences in the San Fernando Valley below (Figure 2.23), a dam failure can be extremely costly in terms of human lives and property damage.

2.2.3 Weak Clay

The problems encountered at soft clay sites include the amplification of the ground motion as well as vigorous soil movement that can damage foundations. Several bridges suffered collapse during the 1989 Loma Prieta Earthquake due to the poor performance of weak clay.

Two parallel bridges were built in 1965 to carry Highway 1 over Struve Slough near Watsonville, CA. Each bridge was 800 ft long with spans ranging from 80 to 120 ft. The superstructures were continuous for several spans with transverse hinges located in spans 6, 11, and 17 on the right bridge and in spans 6, 11, and 16 on the left bridge (they are both 21-span structures). Each bent was composed of four 14-in.-diameter concrete piles extending above the ground into a cap beam acting as an end diaphragm for the superstructure. The surrounding soil was a very soft clay (Figure 2.24). The bridges were retrofit in 1984 by adding cable restrainers to tie the structure together at the transverse hinges.

During the earthquake the soft saturated soil in Struve Slough was violently shaken. The soil pushed against the piles, breaking their connection to the superstructure (Figure 2.25), and pushing them away from the cap beam so that they punctured the bridge deck (Figure 2.26). Investigators arriving at the bridge found shear damage at the top of the piles, indicating that the soil limited the point of fixity of the piles to near the surface. They also found long, oblong holes in the soil, indicating that the piles were dragged from their initial position during the earthquake. It was believed that the damage at Struve

FIGURE 2.14   Gravity retaining wall with shear damage from landslide.

Slough was the result of vertical acceleration, but the structure's vertical period of 0.20 s was too short to be excited by the ground motion at this site.

Similarly, The Cypress Street Viaduct collapsed only at those locations that were underlain by weak Bay mud. This was a very long, two-level structure with a cast-in-place, RC, box girder superstructure with spans of 68 to 90 ft. The substructure was multicolumn bents with many different configurations including some prestressed top bent caps. Most of the bents had pins (shear keys) at the top or bottom of the top columns and all the bents were pinned above the pile caps as well. There was a superstructure hinge at every third span on both superstructures. Design began on the Cypress Viaduct in 1949, and

FIGURE 2.15 Fabric retaining wall damaged during the 1999 Ji Ji, Taiwan, earthquake.

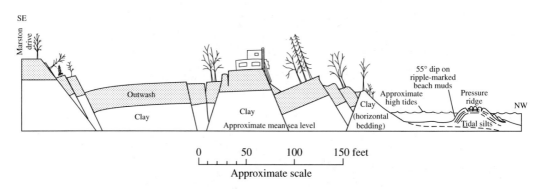

FIGURE 2.16 Section through eastern part of Turnagain Heights Slide. (Drawing courtesy of the National Academy of Sciences. National Research Council (U.S.) no. 1603, National Academy of Sciences, Washington, DC, 1968–1973, 8 v. in 10.)

construction was completed in 1957. The pins and hinges were used to simplify the analysis for this long, complicated structure. The northern two thirds of the Cypress Viaduct was on Bay Mud with 50-ft-long piles while the southern one third was on Merritt sand with 20-ft piles (Figure 2.27).

During the 1989 Loma Prieta Earthquake, the upper deck of the Cypress Viaduct collapsed from Bent 63 in the south all the way to Bent 112 in the north. Only Bents 96 and 97 remained standing. This collapse was the result of the weak pin connections at the base of the columns of the upper frame (Figure 2.28). There was inadequate confinement around the four #10 bars to restrain them during the earthquake. However, the soft Bay mud also played a role in the collapse. The southern portion of the bridge with the same vulnerable details but supported on sand remained standing (Figure 2.29). The northern portion was supported by soft Bay mud that was sensitive to the long-period motion and caused large movements that overstressed the pinned connections.

Buildings on weak clay also are susceptible to earthquake damage. Mexico City was located 350 km from the epicenter of the magnitude 8.1, 1985 Mexico Earthquake, but the city is underlain by an

FIGURE 2.17 Aerial view of Turnagain Slide. (Photograph courtesy of Steinbrugge Collection, Earthquake Engineering Research Center, University of California, Berkeley.)

FIGURE 2.18 About 75 homes were damaged as a result of the Turnagain Heights Slide. (Photograph courtesy of Steinbrugge Collection, Earthquake Engineering Research Center, University of California, Berkeley.)

old lakebed composed of soft silts and clays (Figure 2.30). This material was extremely sensitive to the long-period (about 2 s) ground motion coming from the distant but high-magnitude (8.1) source, as were the many medium-height (10 to 14 story) buildings that were damaged or had collapsed during the earthquake (Figure 2.31). Many much taller and shorter buildings were undamaged due to the difference in their fundamental period of vibration.

FIGURE 2.19   Collapse of Tsu Wei Bridge due to landslide during the Ji Ji, Taiwan, earthquake.

FIGURE 2.20   Landslides at Pacoima Dam following the 1971 San Fernando Earthquake. (Photograph courtesy of Steinbrugge Collection, Earthquake Engineering Research Center, University of California, Berkeley.)

2.3   Damage as a Result of Structural Problems

2.3.1 Foundation Failure

Usually, it is the connection to the foundation or an adjacent member rather than the foundation itself that is damaged during a large earthquake. However, materials that cannot resist lateral forces, such as hollow masonry blocks, make a poor foundation and their use should be avoided (Figure 2.32).

Engineers will occasionally design foundations to rock during earthquakes as a way of dissipating energy and of reducing the demand on the structure. However, when the foundation is too small, it can become unstable and rock over. During the magnitude 7.6, 1999 Ji Ji, Taiwan, earthquake, a local

FIGURE 2.21 Damage to the Lower San Fernando Dam. (Photograph courtesy of Steinbrugge Collection, Earthquake Engineering Research Center, University of California, Berkeley.)

FIGURE 2.22 Closer view of damage to the Lower San Fernando Dam. (Photograph courtesy of Steinbrugge Collection, Earthquake Engineering Research Center, University of California, Berkeley.)

three-span bridge rocked over transversely due to small, drum-shaped footings that provided little lateral stability (Figure 2.33).

We have already seen pile damage as a result of weak clay on the Struve Slough bridges during the 1989 Loma Prieta Earthquake. Similar damage occurred during the 1964 Great Alaska Earthquake. After the 1971 San Fernando and 1995 Kobe Earthquakes, an inspection was made of bridge foundations, but only a little damage to the tops of piles was found. As long as the foundation is embedded in good material, it usually has ample strength and ductility to survive large earthquakes. Usually, it is the more vulnerable

FIGURE 2.23 Aerial view of Lower San Fernando Dam and San Fernando Valley. (Photograph courtesy of Steinbrugge Collection, Earthquake Engineering Research Center, University of California, Berkeley.)

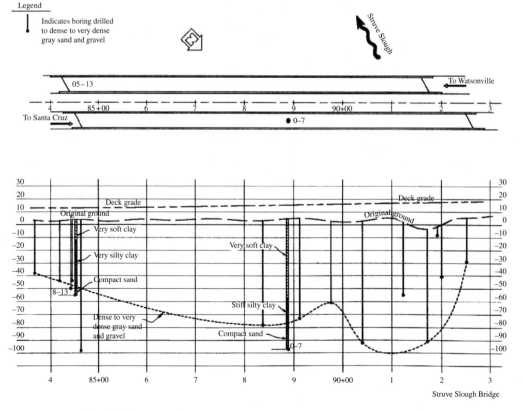

FIGURE 2.24 Soil profile for Struve Slough bridges.

FIGURE 2.25   Broken piles under bridge.

FIGURE 2.26   Piles penetrating bridge deck.

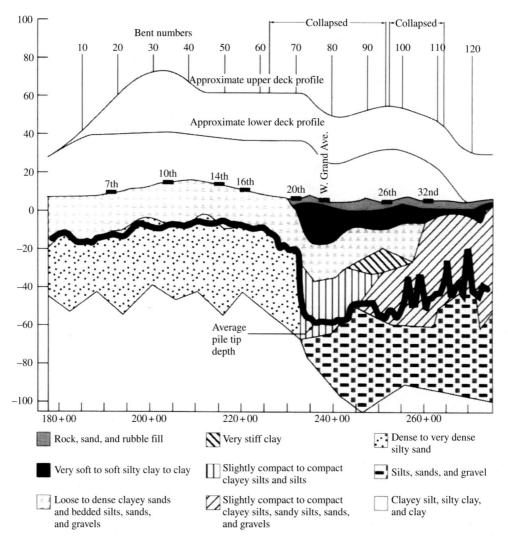

FIGURE 2.27 Geology of Cypress Freeway Viaduct site. (Housner et al., Report to the Governor George Deukmejian, State of California Office of General Services May 1990.)

elements above the foundation that can fail or become damaged during earthquakes. Still, as structures are designed to resist larger and larger earthquakes, we may begin to see more foundation damage.

2.3.2 Foundation Connections

The major cause of damage to electrical transformers, storage bins, and a variety of other structures and lifeline facilities during earthquakes is the lack of a secure connection to the foundation. Houses need to be anchored to the foundation with hold-downs connected to the stud walls and anchor bolts connected to the sill plates. Otherwise, the house will fall off its foundation as shown in Figure 2.34.

The connections to bridge foundations also need to be carefully designed. Route 210/5 Separation and Overhead was a seven-span RC box girder bridge with a hinge at Span 3 and seat-type abutments. The superstructure was 770 ft long on a 800-ft-radius curve. The piers were 4 ft by 6 ft single-column bents. Piers 2 and 3 were on piles, while Piers 4 to 7 were supported by 6-ft-diameter drilled shafts. This interchange was built on consolidated sand.

FIGURE 2.28   Damage to Cypress Street Viaduct.

FIGURE 2.29   Aerial view of Cypress Viaduct showing collapse where the structure crossed over Bay mud.

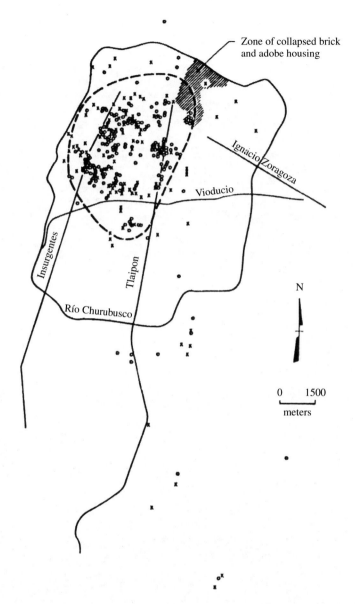

FIGURE 2.30 Locations of building damage at old lake bed in Mexico City (courtesy of EERI; EERI, *Earthquake Spectra*, 4, 3, 569–589, 1988).

During the 1971 San Fernando Earthquake, this structure collapsed onto its west (outer) side, breaking into several pieces and causing considerable damage to two lower-level bridges. A close examination of the fallen structure revealed that the collapse was due to pull-out of the column reinforcement from the foundations.

There was no top mat of reinforcement (and no ties) in the pile caps at Piers 2 and 3. The column longitudinal reinforcement (22 #18 bars) was placed in the footing with 12″ 90° bends at the bottom of the reinforcement. Transverse reinforcement was #4 bars at 12 in. around the longitudinal reinforcement. During the earthquake, the longitudinal reinforcement did not have sufficient development length to transfer the force to the footings. Insufficient confinement reinforcement in the footings and

FIGURE 2.31 Damaged 10-story building between the Plaza de la Constitution and Zona Rosa in Mexico City. (Photo by Karl Steinbrugge, from the EERC NISEE Photo Library.)

FIGURE 2.32 Failed hollow concrete block foundation during the magnitude 6.0, 1987 Whittier, California, earthquake. (Photo by Karl Steinbrugge. Photo courtesy of Steinbrugge Collection, Earthquake Engineering Research Center, University of California, Berkeley.)

columns, and the lack of a top mat of reinforcement resulted in the rebar (and columns) pulling out of the footing (Figure 2.35). Piers 5 to 7 had straight #18 bars embedded 6 ft into pile shafts, and they also pulled cleanly out during the earthquake (Figure 2.36).

After the San Fernando Earthquake, the development length of large-diameter bars was increased, splices to longitudinal rebars were no longer allowed in the plastic hinge area, and more confinement steel was provided in footings and columns.

FIGURE 2.33   Three-span bridge rocked over during the 1999 Ji Ji, Taiwan, earthquake.

FIGURE 2.34   House that fell from its foundation during the 1971 San Fernando Earthquake. (Photograph courtesy of Steinbrugge Collection, Earthquake Engineering Research Center, University of California, Berkeley.)

2.3.3 Soft Story

During the 1989 Loma Prieta Earthquake, many houses and apartment buildings in the San Francisco area had severe damage on the ground floor. These structures had less lateral support on the ground floor to allow room for cars to park under the structure. The remaining supports could not support the movement of the upper stories and dropped the top stories onto the ground (Figure 2.37).

However, a soft story does not always occur on the bottom floor. During the 1995 Kobe, Japan, earthquake, many tall buildings had damage at the midstory, often due to designing the upper floors for a reduced seismic load.

FIGURE 2.35   Failure of column to pile shaft connection.

FIGURE 2.36   Failure of column to footing connection.

Most buildings in Japan are either built of RC or of steel and reinforced concrete (SRC). These SRC buildings, when correctly designed, provide a great deal of ductility and more fire protection during large earthquakes. However, the design practice in Japan was to discontinue either the RC or the SRC above a certain floor. Figure 2.38 shows typical details used in SRC buildings.

Figure 2.39 shows a ten-story SRC building where the third story collapsed during the 1995 Kobe Earthquake.

2.3.4 Torsional Moments

Curved, skewed, and eccentrically supported structures often experience a torsional moment during earthquakes.

FIGURE 2.37   Soft story collapse in San Francisco during the 1989 Loma Prieta Earthquake. (Photo courtesy of the USGS; The Loma Prieta, USGS, 1998.)

A nine-story building in Kobe, Japan, consisted of shear walls along three sides and a moment-resisting frame on the fourth (east) side (Figure 2.40). Shaking during the 1995 Kobe Earthquake caused a torsional moment in the building. The first-story columns on the east side failed in shear, the building leaned to the east (Figure 2.41), and it eventually collapsed (Figure 2.42).

Since rivers, railroad tracks, and other obstacles do not usually cross perpendicularly under bridge alignments, columns and abutments must be built on a skew to accommodate them. These skewed bridges are vulnerable to torsion.

The Gavin Canyon Undercrossing consisted of two bridges over 70 ft tall, with a 67° skew, and was composed of three frames. An integral abutment and a two-column bent supported each end-frame. The center frame was supported by two 2-column bents while supporting the cantilevered end-frames. The superstructure was RC box girders at the end-frames and posttensioned concrete box girders at the center-frame. Each column was a 6 ft by 10 ft rectangular section, fixed at the top and bottom, with a flare at the top. The bridges were retrofitted in 1974 with cable restrainer units at transverse in-span hinges with an 8-in. seat width that connected the frames.

During the 1994 Northridge Earthquake, the superstructures were unseated due to the following factors. The tall, center frame had a large, long-period motion that was out of phase with the stiff end-frames. The end-frame center of stiffness was near the abutment, while its mass was near the bent, causing the end-frame to twist about the abutment. The sharp skew allowed the acute corners to slide off the narrow seats. The cable restrainers, being among the first in the country, were grouted in the ducts, making them too brittle and prone to failure. Both bridges failed as shown in Figure 2.43.

Another interesting example of torsional damage occurred at the Ji Lu Bridge during the 1999 Ji Ji, Taiwan, earthquake. This is a cable-stayed bridge with a single tower and cast-in-place, 102-m-long, box girder spans sitting on two-column end bents that connect the structure to precast "I" girder approach spans (Figure 2.44). The tower is 58 m from the top of the deck to the top of the tower, and 20 m from the top of the footing to the soffit. All the foundations are supported on driven piles. Construction was almost completed on this bridge at the time of the earthquake. All of the cables had been tensioned and all but one had been permanently socketed into the tower. The false work had been pretty much removed except for a few final pours for the portion of the superstructure where it connects to the tower.

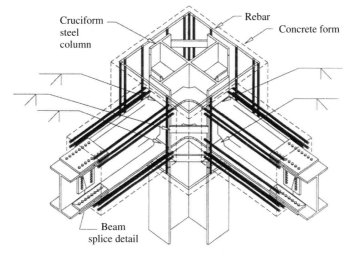

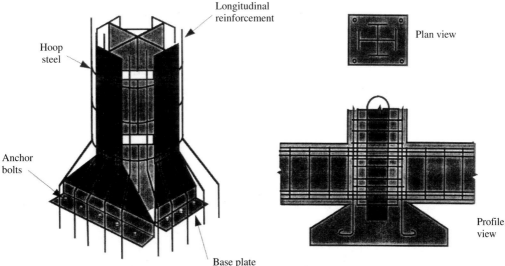

FIGURE 2.38 Examples of SRC construction. (Drawing courtesy of NIST; NIST, The January 17, 1995 Hyogoken (Kobe) Earthquake, U.S. NIST, Gaithersburg, MD, 1969.)

The dominant mode of shaking for this structure was twisting of the tower as the two cantilever spans moved back and forth. Looking at Figure 2.45 we can see that the key at the end of the spans walked up and down the bent seat almost to the end of the support. T.Y. Lin engineers explained that this was because the final pour around the tower had not been completed, making this structure extremely flexible in this direction.

Similar damage occurred to the center piers of curved ramps to the Minatogawa Interchange during the Kobe Earthquake. In this case, the superstructure swung off its end supports and the center column suffered severe torsional damage (Figure 2.46).

As one member of a bridge goes into flexure it can create large torsional moments in adjacent members. For instance, flexure in the columns of outrigger bents causes large torsional moments in the bent cap (Figure 2.47). All of these examples reinforce the idea that most structures require consideration of all three translations and rotations.

FIGURE 2.39 Ten-story SRC building with third floor collapse during the Kobe Earthquake. (Photo courtesy of NIST; NIST, The January 17, 1995 Hyogoken (Kobe) Earthquake, U.S. NIST, Gaithersburg, MD, 1969.)

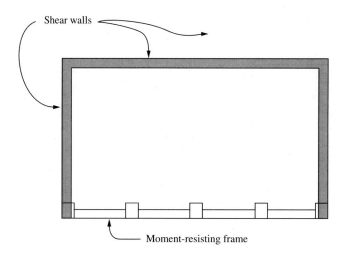

FIGURE 2.40 Plan view of nine-story SRC building in Kobe.

2.3.5 Shear

Most building structures use shear walls or moment-resisting frames to resist lateral forces during earthquakes. Damage to these systems varies from minor cracks to complete collapse. Figure 2.48 is a photo of the Mt. McKinley Apartments after the 1964 Great Alaska Earthquake. It was a 14-story RC building composed of narrow exterior shear walls and spandrel beams (Figure 2.48), as well as interior and exterior columns and a central tower. During the 1964 earthquake, this structure suffered major structural damage to most of the load bearing members.

The most serious damage was to a shear wall on the north side of the building (Figure 2.49). A very wide shear crack split the wall in two directly under a horizontal beam. This crack was because there was not enough transverse reinforcement to hold the wall together as it moved transversely and also due to a cold joint in the concrete at that location. The spandrel beams between the walls had large "X" cracks

FIGURE 2.41 Nine-story SRC building immediately after the 1995 Kobe Earthquake. (Photo courtesy of NIST; NIST, The January 17, 1995 Hyogoken-Nanbu (Kobe) Earthquake, U.S. NIST, Gaithersburg, MD, 1996.)

FIGURE 2.42 Eventual collapse of nine-story SRC building after the Kobe Earthquake. (Photo courtesy of NIST; NIST, The January 17, 1995 Hyogoken-Nanbu (Kobe) Earthquake, U.S. NIST, Gaithersburg, MD, 1996.)

FIGURE 2.43   Damage to Gavin Canyon UC during the 1995 Northridge Earthquake.

FIGURE 2.44   The Ji Lu cable-stayed bridge after the 1999 Ji Ji, Taiwan, earthquake.

FIGURE 2.45   Damage at end-supports to Ji Lu cable-stayed bridge.

FIGURE 2.46   Column damage to the Minatogawa interchange during the Kobe Earthquake.

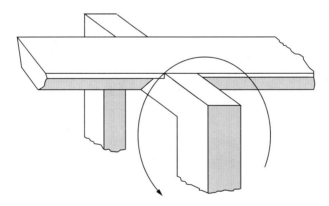

FIGURE 2.47   Column flexure causing torsion in the bent cap.

associated with shear damage as the building moved back and forth. These cracks decreased in size on the upper floors.

There was also shear damage to many of the columns. Figure 2.50 is one of the exterior columns on the south side of the building with a diagonal shear crack. Again, the problem was insufficient transverse reinforcement to resist the large shear forces that occurred during the earthquake. The central tower was also damaged. However, the Mt. McKinley Apartment can be viewed as a success since there was no collapse and no lives lost during this extremely large earthquake.

Bridges are equally susceptible to shear damage. For instance, there was considerable shear damage to piers on elevated Route 3 during the Kobe Earthquake. The superstructure is mostly steel girders and the substructure is RC single-column bents. Between Piers 148 and 150, the superstructure is a three-span, continuous, double-steel box with span lengths of 45 m, 75 m (between Piers 149 and 150), and 45 m. Pier 149 is a 10-m-tall by 3.5-m square RC single-column bent supported on a 12-m square pile cap.

FIGURE 2.48 West elevation of the Mt. McKinley Apartment building after the 1964 Great Alaska Earthquake. (Photograph by Karl Steinbrugge, Earthquake Engineering Research Center, University of California, Berkeley.)

FIGURE 2.49 Damage to north side of Mt. McKinley Apartments. (Photograph by Karl Steinbrugge, Earthquake Engineering Research Center, University of California, Berkeley.)

Pier 150 is a 9.1-m-tall by 3.5-m square RC single-column bent supported on a 14.5-m by 12-m rectangular pile cap.

Figure 2.51 shows the shear failure at Pier 150. This damage was the result of insufficient transverse reinforcement and poor details. During the large initial jolt (amplified by near-field directivity effects) the transverse reinforcement came apart and resulted in the column failing in shear. Pier 149 was also severely damaged, but because it was taller (and more flexible), most of the force went to stiff Pier 150. The three-span continuous superstructure survived the collapse of Pier 150 with minor damage.

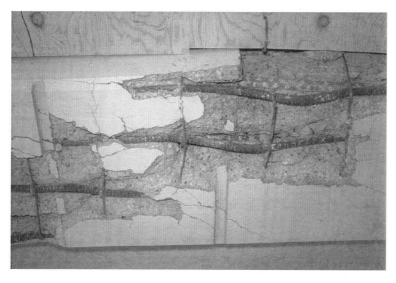

FIGURE 2.50 Damage to the south side of Mt. McKinley Apartments. (Photograph by Karl Steinbrugge, Earthquake Engineering Research Center, University of California, Berkeley.)

FIGURE 2.51 Shear failure of Pier 150 on Kobe Route 3.

2.3.6 Flexural Failure

Flexural members are often designed to form plastic hinges during large earthquakes. A plastic hinge allows a member to yield and deform while continuing to support its load. However, when there is insufficient confinement for RC members (and insufficient *b/t* ratios for SRC members), a flexural failure will occur instead. Often, flexural damage is accompanied by compression or shear damage as the capacity of the damaged area has been lowered.

The Dakkai subway station in Japan is a two-story underground RC structure. It was constructed by removing the ground, building the structure, and then covering it (the-cut-and-cover method). During the 1995 Kobe Earthquake, the center columns on both levels suffered a combination of flexural and compression damage that caused both roofs to collapse along with a roadway that ran above the station (Figure 2.52). Figure 2.53 shows the rather slender center columns at the lower level after the earthquake. The columns had insufficient transverse reinforcement at the location of maximum moment. The transverse reinforcement broke as the columns were displaced, allowing the longitudinal reinforcement to buckle and the concrete to fall out of the column.

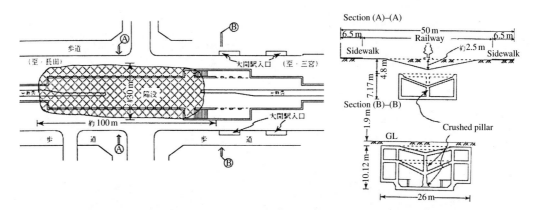

FIGURE 2.52 Plan and section drawings of the Dakkai subway station after the Kobe Earthquake. (Drawing courtesy of JSCE, JSCE, Preliminary report on the Great Hanshin Earthquake, January 17, 1995, JSCE, 1995.)

FIGURE 2.53 Flexural damage to columns at lower level of Dakkai subway during the 1995 Kobe Earthquake. (Photo courtesy of JSCE, JSCE, Preliminary report on the Great Hanshin earthquake, January 17, 1995, JSCE, 1995.)

Steel columns experience similar damage when the flexural demand exceeds the capacity. In downtown Kobe, in the Nagata District, Route 3 splits into two parallel structures with the super-structure composed of 50-m-long simple spans with steel girders and a three-span, continuous, steel box girder section between Piers 585 and 588. The substructure consists of 2.2-m-diameter, 14-m-tall steel hammerhead single columns. The steel columns are bolted onto 4-m-diameter hollow, concrete 20-m-long shafts. The column bottoms are filled with concrete to protect against vehicular impact. During the Kobe Earthquake, these steel columns had damage varying from local buckles to a complete section buckle, and at a few locations, the steel shell had torn, splitting the column. Most of the buckling occurred in a thinner section of the column. In some cases (Figure 2.54), the column underwent an excursion in only one direction and consequently had a buckle on one side of the column.

In some cases the buckled face tore in the tension cycle (Figure 2.55). The tears occurred in low-ductility welds. After the earthquake, the columns were tilting dangerously to the side. Buckling occurred before a plastic hinge was formed. A few columns remained undamaged as a result of failed bearings. Although local buckling cannot be completely eliminated, its spread can be prevented by maintaining smaller *b/t* ratios. This is accomplished with thicker sections, more frequent stiffeners, and diaphragms, or by filling the steel shells with lightweight concrete.

FIGURE 2.54 Pier 585 on Kobe Route 3 during the 1995 Kobe Earthquake. (Photo courtesy of Hanshin Expressway Public Corporation.)

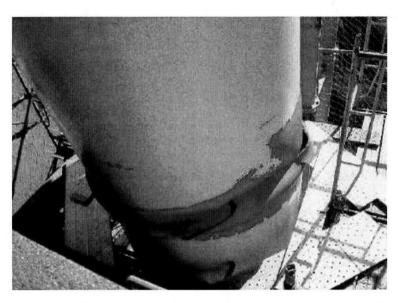

FIGURE 2.55 Torn buckle of steel column on Route 3 after the 1995 Kobe Earthquake. (Photo courtesy of Hanshin Expressway Public Corporation.)

2.3.7 Connection Problems

The most catastrophic type of structural damage is the failure of connections. When a bridge superstructure moves off its expansion joint or when the connection between building columns and beams fail, the result is too often the collapse of the structure.

Between Piers 39 and 43 on Kobe Route 3, the superstructure was a series of 52-m-long simple-span steel boxes supporting a concrete deck. Steel web plate restrainers with oversized holes at one end connected the girders together over each pier support. Fixed pin bearings and movable roller bearings supported each span. The substructure was hammerhead piers with 3.5-m-diameter circular or 3.5-m square concrete columns over 10 m in height. Each column was supported on a rectangular footing supported by 18- to 1.0-m-diameter by 16.5-m-long piles.

During the earthquake, the piers moved back and forth longitudinally, sustaining cracks or shear damage at the column bases. At the same time, the girder spans were moving west (toward Pier 43) as a result of the impact with the five-span, continuous, steel girder bridge east of Pier 38. The relative displacement between the piers and the simple spans exceeded the 0.8-m seat width at Piers 40 and 41 and the expansion end of Spans 40 and 41 fell off the piers (Figure 2.56). Most of the web plate restrainers failed in tension (Figure 2.57) and almost all of the fixed and expansion bearings were damaged, mostly due to the top shoe pulling out of the bearing (Figure 2.58).

RC structures must be carefully designed to allow the shear transfer at joints. A common kind of residence in Turkey was four- to eight-story buildings composed of concrete columns supporting concrete slabs and infill walls of unreinforced masonry blocks. During the 1999 Kocaeli Earthquake, these buildings collapsed at very low levels of acceleration, killing thousands of people. The large inertia force from these heavy structures had to be carried by the slender columns and by the inadequately reinforced connections (Figure 2.59). Sufficient concrete and reinforcement must be provided to resist the large tension and compression forces in moment-resisting joints during earthquakes. Joints of moment-resisting RC frames should be stronger than the elements that join them.

The connections of steel moment-resisting frames, with detailing recommended by design codes prior to 1994, suffered considerable damage during the Northridge (and Kobe) Earthquakes. During the Northridge Earthquake, over 100 steel moment-resisting frame buildings had some damage to the

FIGURE 2.56   Superstructure collapse at Spans 40 and 41 on Kobe Route 3.

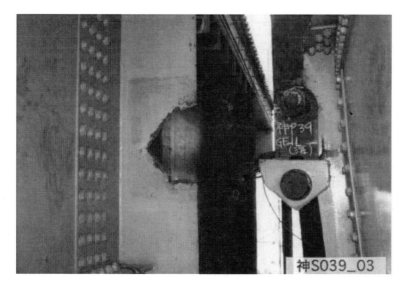

FIGURE 2.57   Broken restrainer at Pier 39. (Photo courtesy of Hanshin Expressway Public Corporation.)

column to beam connections. As previously mentioned, flexural elements are supposed to form plastic hinges during large earthquakes. However, during the Northridge Earthquake the connection fractured instead. In this connection the beam flanges are welded to the column with a column flange stiffener placed along the column web to provide continuity to the joint. The beam web is bolted to the column with a shear connection plate. As shown in Figure 2.60, a large crack fractured the flange and web of some columns. Although this damage was unexpected and looked serious, no building during the Northridge Earthquake was reported to have collapsed as a result of the failure of this connection. Still, a few more cycles of motion (from a much larger earthquake than the magnitude 6.7 event) could easily

FIGURE 2.58   Broken bearings at Pier 40. (Photo courtesy of Hanshin Expressway Public Corporation.)

FIGURE 2.59   Collapsed reinforced concrete building during the 1999 Kocaeli Earthquake.

tear apart this fractured column. Consequently, new connection details have been developed as well as retrofit details for the many buildings with this connection.

2.3.8 Problem Structures

Some types of structures have performed particularly poorly during previous earthquakes. Usually, this is the result of vulnerabilities such as weak connections, improper detailing, and eccentric loads that predispose these structures to severe damage and collapse during earthquakes.

Unreinforced masonry should never be used to resist lateral forces since it is very weak in tension and very heavy, resulting in walls that immediately fall over, seriously injuring anyone nearby (Figure 2.61). However, reinforced masonry walls have performed very well when thoughtfully designed.

FIGURE 2.60   Steel moment-resisting frame connection after the 1994 Northridge Earthquake. (Photo courtesy of EQE International; EQE, *EQE Int. Rev.*, Fall, 1–6, 1994.)

FIGURE 2.61   Rear view of damaged unreinforced masonry (URM) building facing Pacific Garden Mall, Santa Cruz Area, California. (Photo by James R. Blacklock, Loma Prieta Collection, Earthquake Engineering Research Center, University of California, Berkeley.)

Tilt-up buildings have also performed poorly during past earthquakes because the walls pull away from the roof diaphragm, because of the discontinuity at the vertical joints between panels, because of poor connections to the roof joists, etc. Similarly, precast prestressed bridges tend to fall apart due to inadequate connections between members.

Tanks, bins, silos, grain elevators, concrete mix plants, etc. are the most commonly damaged structures during earthquakes. They are tall, heavy, and too often designed with weak supports and inadequate anchors (Figure 2.62).

Construction sites are particularly dangerous places to be on during an earthquake. Too often, little thought is given to providing lateral strength to partially constructed structures. This can result in millions of dollars in damage as well as fatalities. For instance a new expressway was being built during the 1999 Ji Ji, Taiwan, earthquake. This included two simple-span, precast, and prestressed "I" girder bridges on two-column bents near the epicenter. The girders were 84 in. tall by 24 in. wide and sat on

FIGURE 2.62 California Water Service 150,000-gal tank on six legs; the typical failure of tanks is upside down, with riser on top (Bakersfield, California, photo by Karl V. Steinbrugge, Steinbrugge Collection, slide/image No. S64, photo date July 25, 1952, Kern County, California, earthquake, July 21, 1952, magnitude 7.69; courtesy of Steinbrugge Collection, Earthquake Engineering Research Center, University of California, Berkeley).

FIGURE 2.63 Collapsed precast, prestressed "I" girders during construction of new expressway in Taiwan due to the 1999 Ji Ji, Taiwan, earthquake.

24 in. by 16 in. by 5 in.-tall elastomeric pads. The bents were at least 30 ft tall from the ground to the top of the bent caps. The bridge had 13 approximately 150-ft-long spans. At the time of the earthquake, the girders had been placed on the bents and the seat-type abutments, and the intermediate and end-diaphragms were just beginning to be cast between the girders. Wherever the intermediate diaphragms had not been placed, the girders had been shaken off of their supports and had fallen to the ground, breaking into pieces (Figure 2.63). This resulted in the loss of about 100 girders. The use of temporary supports before the casting of diaphragms could have saved about a million dollars.

All of these structures can be designed to perform adequately when they are provided with sufficient strength and ductility and a good understanding of their structural behavior during earthquakes.

2.4 Secondary Causes of Structural Damage

2.4.1 Surface Faulting

Few structures are designed to accommodate an offset of several meters laterally or vertically at any location. Yet, that is what is required near active faults. When the fault reaches the surface, the ground can be pushed together, pulled apart, raised, or dropped 5, 6, and even 10 m. Some of the most spectacular structural damage from faulting occurred during the 1999 Ji Ji, Taiwan, earthquake. The Shih Kang Dam was one of several structures across the ill-fated Dajia River north of Route 3 (Figure 2.64). During the earthquake, the north end of the dam dropped 9 m (Figure 2.65).

The Dajia River Bridge was a 14-span, RC bridge with a "T" girder superstructure, single-column bents with hammerhead bent caps, and drum-shaped footings. The girders sat on elastomeric pads and between transverse shear keys. It was one of several bridges a little north of Route 3 across the Dajia River near the town of Shargang (Figure 2.64).

This bridge had spans 30 m long by 10 m wide and columns about 10 m in height at the center of the bridge. During the earthquake, the first three southerly spans collapsed along with Pier 3 (Figure 2.66). The south abutment and first two piers moved 6.5 m vertically and 3 m to the west. The column

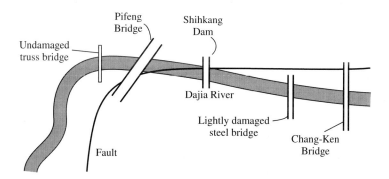

FIGURE 2.64   Location map of damaged structures along the Dajia river after the 1999 Ji Ji, Taiwan, earthquake.

FIGURE 2.65   North end of Shih Kang Dam after the Ji Ji, Taiwan, earthquake.

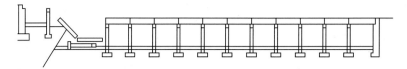

FIGURE 2.66   Elevation drawing of Dajia Bridge collapse.

FIGURE 2.67   The Dajia River bridge and 6-m fault scarp.

foundation rotated out of ground. The bridge collapse was due to a 6-m fault scarp that created a waterfall alongside the bridge (Figure 2.67).

2.4.2   Damage Caused by Nearby Structures and Lifelines

Weak or poorly designed facilities will often cause damage to nearby structures during earthquakes. The intelligent planning of adjacent lifelines (called colocation) is a good way to minimize disruptions to lifelines during earthquakes and other disasters.

When buildings, bridges, and other structures are built too closely together, one problem is pounding. This is particularly a problem for connectors on highway interchanges that are sometimes in such close proximity that the column of a higher structure will go through the lower bridge deck. Buildings are often built too closely together causing pounding and moderate structural damage during earthquakes. Pounding can cause serious damage since impact loads of very short duration can carry a very large force. However, pounding may also be beneficial when it prevents resonance.

More serious damage occurs when a poorly designed structure collapses onto an otherwise seismically resistant structure. There are no benefits from this interaction.

Piers 352 and 353 on the Route 3 Expressway in Kobe, Japan, supported the superstructure on steel hammerhead piers. The columns were 2 m by 1.5 m by 8.8 m in height, supported on pile foundations, and built using the 1964 specifications. The piers were 8.8 m tall and on pile foundations. The superstructure consisted of 30-m-long, steel simply supported plate girders, one end fixed and the other end on elastomeric expansion bearings. During the 1995 Kobe Earthquake, the bridge was damaged when a collapsing building fell against the expressway. The building (shown in Figure 2.68) applied a lateral load to the superstructure. This resulted in buckling of the vertical stiffeners on the girder webs, bearing damage, and also some damage to the bottom of Pier 353. The main damage was separation of a metal plate that was installed to protect the steel column from traffic collisions. However, because there was a collapsed building leaning against the bridge, it was felt to be expedient to provide shoring to the piers after the earthquake. The ability of the superstructure to move laterally over the piers helped to prevent more serious damage from occurring to the substructure.

FIGURE 2.68 Collapsed building leaning against the Kobe Route 3 Expressway. (Photo courtesy of Hanshin Expressway Public Corporation.)

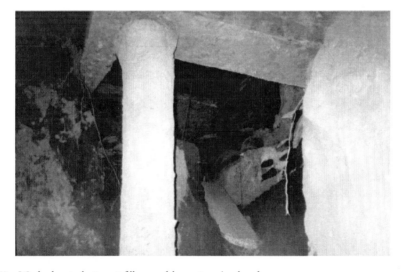

FIGURE 2.69 Washed-out abutment fill caused by water pipe break.

A lifeline can often cause problems to other nearby lifelines during earthquakes. The Balboa Boulevard Overcrossing is a three-span (actually two spans with a 22-ft-long bin-type abutment between Abutment 1 and Bent 2) cast-in-place prestressed concrete bridge (except for a concrete slab for the bin-type abutment) on Highway 118 near Northridge, CA. The superstructure is 283 ft long and 117 ft wide. Bent 3 is a three-column bent on a spread footing. The columns are 5-ft octagonal sections with a large flare. The abutments and wingwalls are on piles. During the 1994 Northridge Earthquake the Balboa Boulevard Overcrossing experienced minor spalling of the concrete cover at the top of the columns during the earthquake. The shaking broke a small-diameter water line under Abutment 1 that washed out the backfill behind the abutment. Figure 2.69 shows the collapsed slab for the bin-type abutment and the cast-in-drilled-hole (CIDH) pile supporting Abutment 1. The grade beam that supports the wingwall is on the left.

These are just a few of the many secondary causes of damage during large earthquakes. What typifies this kind of damage is that its prevention may not rely on structural issues. In fact, the best solution is to find a better site during the planning of the structure. There is little the engineer can do once the structure straddles an active fault or is within a few feet of a vulnerable structure.

2.5 Recent Improvements in Earthquake Performance

Much of the damage that has occurred during recent earthquakes was the result of the failure of the surrounding soil or structural elements. Not surprisingly, most of the recent improvements to the earthquake resistance of structures have focused on methods of improving poor soil, on soil–structure interaction, and on structural elements and systems that increase strength, stability, ductility, etc.

There has also been some work to prevent secondary earthquake damage. For instance, California has begun to retrofit bridges that cross over active faults. On the Colton Interchange, a bridge was provided with additional supports to catch the superstructure if it is pulled off its piers. Another California bridge was provided with large galleries at the abutments to accommodate large movements. A couple of bridges have even been provided with gates that close the bridge when the ground begins to shake.

Some effort has also been made to improve lifeline performance, most notably in Wellington, New Zealand, where facilities are carefully planned and located to prevent disruptions in service during large earthquakes.

2.5.1 Soil Remediation Procedures

Because much of Japan is covered with either weak clay or saturated, loosely consolidated material, the development of soil remediation procedures has flourished in that country. After the 1994 Kobe Earthquake researchers did a careful study of soil-remediation sites and found that these locations performed much better than the surrounding area [10].

The following are a few examples of soil improvement methods commonly used in Japan.

2.5.1.1 Gravel Drains: Ariake Quay-Wall Improvement Project

Ariake is a man-made island in Tokyo Bay. The weak material of the island is supported by a quay-wall made of timber piles and steel sheet piles. The goal of this project was to protect the quay-wall from damage during an earthquake (Figure 2.70). The wharf is composed of loose sand that could liquefy during an earthquake. Increased pore water pressure would then push over the quay-wall. Gravel drains were installed to reduce pore water pressure during an earthquake. The advantage of this method is that it is quick, inexpensive, and free of vibration. The disadvantage is that it does not prevent settlement of

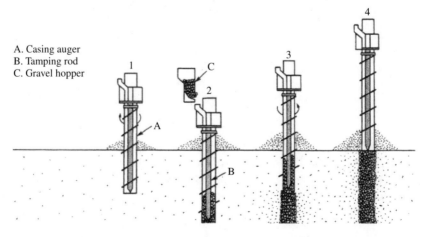

A. Casing auger
B. Tamping rod
C. Gravel hopper

FIGURE 2.70   Gravel drain construction. (Drawing courtesy of Japan's Public Works Research Institute.)

the soil during an earthquake. Since the goal is protection of the quay-wall, this is not considered a problem. The procedure is to drill a hole in the ground using a casing auger. Gravel is carried to the auger by a front-end loader. It is dropped into a hopper, lifted to the top of the auger, and poured into the casing. The casing is then removed from the ground leaving a sand drain. The gravel is fairly uniform. The casing has replaceable steel teeth to help it cut through soil and push away rocks. The gravel drains are placed close enough together to form a grid that will effectively drain out all the water. This project consisted of 3997 gravel drain piles in an area of 2770 m^2 for 1.4 piles/m^2 (the drains are 0.8 m apart). The drains are 0.5 m in diameter and 17 m long and cost about 50,000 Yen ($500) per pile or about 200 million Yen for the project. It takes less than an hour to make a gravel drain and there were four augers at the job site.

2.5.1.2 Deep Mixing Method: Kawaguchi City

The area along the Arakawa River has poorly consolidated soil. This project used deep soil mixing to improve the unconfined compressive strength of the ground (Figure 2.71). Without the DMM the soil would be expected to settle 1.5 m after the embankment is placed. With the DMM the settlement should be less than 0.03 m.

On this project, a modified pile-driving machine rotates a pair of rods with stirring wings for mixing the soil. The distance between piles can be changed by moving the rods closer together or further apart. The pile diameter is 1 m on this project. First, the machine pushes the stirring wings down 30 m,

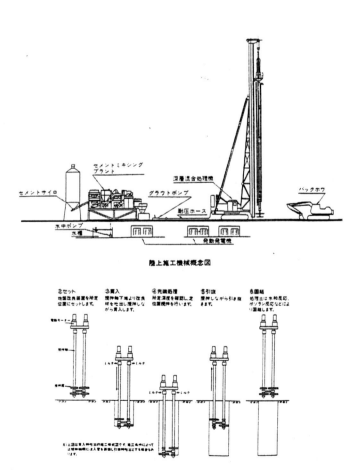

FIGURE 2.71 Soil deep mixing construction. (Drawing courtesy of Japan's Public Works Research Institute.)

breaking up the soil and making it soft and permeable. The wings rotate 20 cycles per minute as they descend into the soil. Then, the stirring wings are pulled up while injecting cement milk into the soil. The milk is composed of an equal weight of cement and water. About 100 to 150 kg of cement milk is injected per cubic yard of soil. The wings rotate at 40 cycles per minute as they ascend. Loose soil comes up during this procedure and is removed with an excavator. The stirring wings descend and ascend at 1 m/min. There is a cement plant for each machine and a separate pump for each rod. The cement milk flows through flexible hoses from the cement plant to the top of the rods where it is injected into the soil. When the piles are completed the machine crawls to the next location on a plywood mat and drills another pile. Display panels in the machine cab give information about depth, rotation, and the amount of cement milk being pumped.

The deep mixing method is also very effective in preventing soil liquefaction. In fact, this method was used to prevent liquefaction for some flood prevention works next to this site. In this case the piles are placed in a lattice pattern to contain the liquefiable soil.

The deep mixing method was found to be the most effective method for creating strong, highly ductile ground that does not liquefy or settle. It is also the most expensive method, but the lack of noise and vibration makes it ideal for city environments.

2.5.1.3 Sand Compaction Pile Method: Ohgishima Island, Tokyo

Sand compaction piles are a popular way of preventing liquefaction of loose alluvium. However, the noise and vibration make it unacceptable at some locations.

This project was on a man-made island in Tokyo Bay (Figure 2.72). It is the location of an LNG tank farm. Soil remediation was required at the toe of an embankment that covers these tanks.

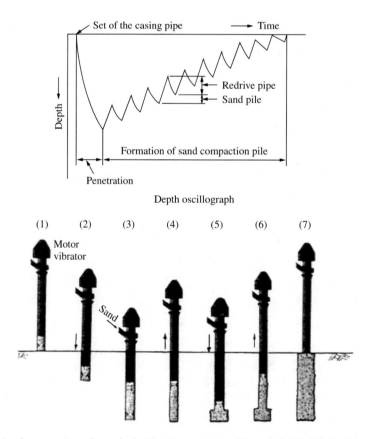

FIGURE 2.72 Sand compaction pile method. (Drawing courtesy of Japan's Public Works Research Institute.)

Because it was a remote site, the loud noise and vibrations were not a problem. The sand compaction pile method uses a modified pile-driving machine to vibrate a steel pipe into the ground. When the penetration reaches the proper depth, sand is carried to the top by a hopper and forced to the bottom of the pipe with compressed air. Then, the pipe is raised and lowered in the hole as sand is repeatedly shot to the bottom of the hole. The result is a pile of compacted sand and an area between the piles of compacted soil. The equipment is similar to the gravel drain method. The pile driver has a steel casing with a lid that can be opened and closed at the bottom. Sand is brought in by dump truck to the site. A front-end loader pours the sand into the hopper. The hopper carries the sand to the top of the pipe where it is poured into the air compression chamber. The steel casing with the lid closed is vibrated to the required depth. In this project the depth was 17 m and the pile diameter was 0.7 m. When the pile is completed, the panels the truck sits on are moved back and the pile driver is moved back to drive a new pile. The pile is driven at about 2 min/m and a pile is completed in about 30 min. Before this project began, the soil had an average N value of 10 to 12. The areas that have been completed now have an N value of 15 to 20.

2.5.2 Improving Slope Stability and Preventing Landslides

The failure of many retaining walls during the 1999 Ji Ji, Taiwan, earthquake was surprising. Well-designed cantilever and MSE walls have performed well during recent earthquakes. For instance, there was an MSE wall that continued supporting the embankment adjacent to the Arifiye Bridge that collapsed during the 1999 Kocaeli, Turkey, earthquake (Figure 2.73). A cribwall in the Santa Cruz Mountains was severely distressed during the 1989 Loma Prieta Earthquake but continued to support the steeply sloping ground (Figure 2.74). There is no record of a cantilever retaining wall failure during a California Earthquake. Therefore, well-designed retaining walls continue to be a good method for supporting most soils.

Other methods of preventing soil movement include planting trees and other vegetation, the use of geotextiles, installing piles, etc. The foundations of structures may also be strengthened to prevent their movement. Many of the methods used for preventing liquefaction can also be used to stabilize soils. However, when an area has a long history of landslides, it may be more prudent to build elsewhere.

FIGURE 2.73 Mechanically stabilized earth wall continues to support the embankment behind the collapsed Arifiye Overcrossing after the 1999 Kocaeli, Turkey, earthquake.

FIGURE 2.74 Crib wall continues to support steeply sloping soil in Santa Cruz after the 1989 Loma Prieta Earthquake. (Photo courtesy of EERI.)

2.5.3 Soil–Structure Interaction to Improve Earthquake Response

Soil–structure interaction modifies the ground-input motion at the foundation. By taking advantage of soil–structure interaction, structures can be protected during earthquakes. Frank Lloyd Wright's 1915 Imperial Tokyo Hotel survived the 1923 Great Kanto Earthquake because he built his foundation in 70 ft of weak clay. During the earthquake, the ground underneath moved violently while his hotel (effectively isolated by this plastic material) remained relatively immobile.

Similarly, a common retrofit technique for bridges is to allow the foundation to rock during large earthquakes. As the foundation rocks, the period is lengthened and damping is increased, all of which lowers the demand on the structure. To insure stability, the foundation may be connected to flexible anchors or an outer perimeter of piles may be placed under (but not connected to) a widened foundation. The piles will provide support for the foundation as it rocks back and forth.

Special foundations are sometimes used to improve seismic response. For instance, a popular bridge foundation in California is the large-diameter drilled pile shaft. These foundations are very flexible, replacing potentially damaging seismic forces with large displacements. Moreover, when the pile shaft is allowed to yield, a large plastic hinge forms, which provides more ductility for the structure.

In contrast to California's efforts to provide more flexibility and ductility in its foundations, Japan has been developing stiffer and more massive foundations for bridges and other structures. Besides the advantage of handling very large forces elastically, many of these new foundations use advanced automation techniques to simplify their construction. *The open caisson construction method* pushes a precast hollow cylindrical caisson into the ground while excavating the ground beneath with a grab bucket. Additional sections are attached to the top until it bears on good material. The caisson segments can be constructed up to 4 m in diameter and are match-cast for a tight fit. After they are completely assembled,

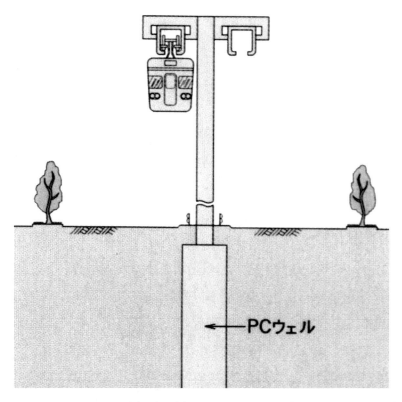

FIGURE 2.75 Precast caissons were used for the Chiba City, Japan, monorail. (Drawing Courtesy of Japan's Public Works Research Institute.)

they are filled with soil, and a steel assembly is attached to mount the substructure. This method was used for the Chiba City Monorail System. The open caisson construction method allowed them to build large caissons often within a few feet of an existing building (Figure 2.75).

Another innovative foundation in Japan is the *continuous diaphragm wall*. This method involves excavating wall-type ditches and casting wall elements in them (Figure 2.76). These walls are connected together with joint elements. Then, an upper slab connects the top of the walls to the pier. The walls typically vary in thickness from 1 to 2.8 m. Continuous diaphragm walls come in a variety of shapes. Circles, rectangles, and grids are all very common. These are used for deep, stiff foundations as well as enormous shafts and storage tanks.

2.5.4 Structural Elements that Prevent Damage and Improve Dynamic Response

Improved structural performance during large earthquakes depends on a balanced structural system. Elements that share the same displacement during large earthquakes must be designed to have about the same stiffness. Otherwise, the stiffer elements will be forced to resist most of the earthquake force. Elements that share the same force are often provided with a fuse to limit the force and protect adjacent members.

Providing great strength to resist earthquakes is usually self-defeating. As the elements are made stronger, they attract larger earthquake forces. If an element along the load path cannot resist these forces, it will break, sometimes with disastrous consequences. Elements like shear walls are often used to limit displacements in buildings, but they are nonductile and will often shatter when unexpectedly large earthquakes occur.

先行エレメント

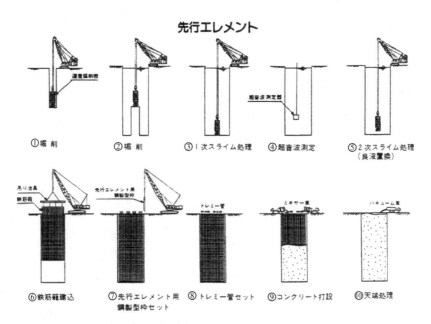

FIGURE 2.76 Construction sequence for diaphragm foundations. (Drawing courtesy of Japan's Public Works Research Institute.)

FIGURE 2.77 Construction of the new San Francisco Airport International Terminal. (Photo courtesy of Earthquake Protection Systems Inc.)

The use of isolation and damping devices has gained popularity because they can protect a structure during several large earthquakes without suffering damage or requiring replacement. Moreover, these devices have proven to be effective for new construction as well as for retrofitting vulnerable existing structures.

The new San Francisco Airport International Terminal is a large steel frame building with a truss roof (Figure 2.77). It is the largest base-isolated building in the world. To ensure that it would remain in service after a very large (magnitude 8) earthquake, 267 friction pendulum seismic isolation bearings were placed between the steel columns and the foundations (Figure 2.78 and Figure 2.79). Each bearing can move 20 in. while supporting 6000 kip. The bearings increased the building's fundamental period to 3 s and reduced the earthquake force by 70%.

These interesting devices are surprisingly simple in principle. For service level loads, static friction restrains the system. During an earthquake, the system operates like a pendulum (Figure 2.80) with the amount of damping controlled by dynamic friction and with the period

FIGURE 2.78 Layout of foundations and friction pendulum devices for the new San Francisco Airport International Terminal. (Photo courtesy of Earthquake Protection Systems Inc.)

FIGURE 2.79 One of the 267 friction pendulum devices for the new San Francisco Airport International Terminal. (Photo courtesy of Earthquake Protection Systems Inc.)

and stiffness as shown below (where R is the radius, W is the weight, and g is the acceleration due to gravity).

$$T = 2\pi\sqrt{\frac{R}{g}} \tag{2.3}$$

$$K = \frac{W}{R} \tag{2.4}$$

The period of the structure is increased by flattening the bearing's concave surface and the force–displacement relationship is modified by changing the friction or the dead load. Similarly, the design of the pyramid-shaped "Money Store" in West Sacramento (Figure 2.81) included fluid viscous dampers (FVD) to absorb energy and allow the steel moment-resisting frame to remain elastic during large

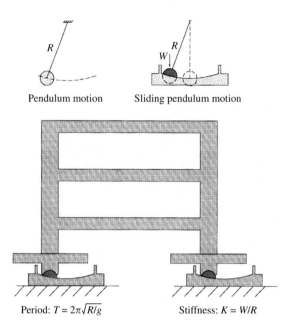

Pendulum motion   Sliding pendulum motion

Period: $T = 2\pi\sqrt{R/g}$ Stiffness: $K = W/R$

FIGURE 2.80 The structure's period is controlled by the radius of the curved bearing. (Drawing courtesy of Earthquake Protection Systems Inc. Zayas et al., The FPS earthquake resisting system, University of California, Berkeley, CA, 1987.)

FIGURE 2.81 The uniquely shaped "Money Store" was designed with fluid viscous dampers to improve its seismic behavior. (Photo courtesy of Marr-Shaffer & Miyamoto, Structural Engineers.)

earthquakes. The FVDs are cylinders filled with liquid silicon. When a load is applied, a piston pushes through the viscous fluid in the cylinder, absorbing energy (Figure 2.82). Concentrically braced frames were built with FVDs between the diagonal braces and the columns to increase the damping ratio of the building to 15% and reduce the story drift ratio to 0.005 (Figure 2.83).

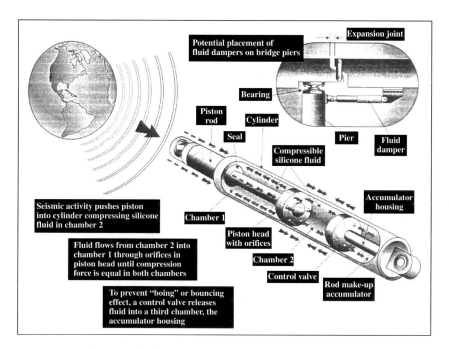

FIGURE 2.82 Schematic drawing of a fluid viscous damper. (Drawing courtesy of Taylor Devices Inc.)

FIGURE 2.83 The steel frames included fluid viscous dampers to reduce the story drift. (Photo courtesy of Marr-Shaffer & Miyamoto, Structural Engineers.)

Isolation and damping devices are equally adept at providing protection for seismically vulnerable existing buildings. The Long Beach VA Hospital is a concrete shear-wall structure that was found to be vulnerable to large earthquakes (Figure 2.84). Many retrofit strategies were studied before isolation was finally accepted as the best way to keep the hospital functioning after a large earthquake on the nearby Pales Verdes and San Andreas Faults. Because the hospital also had to remain in service during construction, the sequence of the construction was crucial and the contractor had to be responsive to any problems that developed that could impact the daily operations of the hospital. Most challenging was completely isolating the structure. A moat had to be dug around the building and flexible connections had to be designed for all the utility lines as they entered the building.

Lead-rubber bearings were installed a few feet above the base of all 150 concrete columns that supported the building (Figure 2.85). Each bearing was 22 in. tall and 24 in. in diameter. First, friction gripping devices were used to transfer the load from the column onto hydraulic jacks. Then, a 20-ft section of the column was replaced with a lead-rubber isolation bearing. These bearings have a lead core that provides an initial high stiffness for service loads. During large earthquakes the lead core yields, providing damping for the structure. The bearings used on the Long Beach VA Hospital will allow the ground to move 16 in. without impacting the building.

Isolation and damping devices can also be used on other structures. The All American Canal Bridge in California is one of the many bridges that uses lead-rubber bearings to isolate the superstructure from earthquake ground motions (Figure 2.86).

The Vincent Thomas Suspension Bridge (a mile from the Long Beach VA Hospital) was retrofit with 80 FVDs to absorb energy and prevent the bridge deck from pounding against the towers and cable bents (Figure 2.87). The connection between the towers and the truss was modified to allow very large relative movements. The truss section on each side of the towers was replaced with a new unit that includes a deck section with 26-ft-long finger joints and large viscous dampers to absorb energy and prevent the truss from pounding against the towers (Figure 2.88).

In general, isolation and damping devices have performed very well during large earthquakes. However, isolated structures have usually been too far away to experience really large accelerations. The one occurrence where isolation and damping devices were very close to a fault rupture was on the Bolu

FIGURE 2.84 The existing Long Beach VA Hospital was seismically retrofit with lead-rubber bearings. (Photo courtesy of Dynamic Isolation Systems, Inc.)

FIGURE 2.85 The concrete columns were cut at midheight and lead-rubber isolation devices were installed. (Photo courtesy of Dynamic Isolation Systems Inc.)

FIGURE 2.86 Lead-rubber isolation bearings on the All American Canal Bridge in California.

FIGURE 2.87 The Vincent Thomas Bridge is a three-span suspension structure built over the port of Los Angeles.

FIGURE 2.88 The Vincent Thomas Bridge retrofit provided gaps between the decks and the towers and also fluid viscous dampers to dissipate energy. (Photo courtesy of Enidine Inc.)

Viaduct during the 1999 Duzce Earthquake. However, in this installation the bearings were eccentric to the dampers, resulting in out-of-plane motions that locked up the dampers and caused significant damage to the bridge. It will probably take several earthquakes to work out all the bugs and come up with installations that work most effectively. Because of the increased use of isolation and damping devices on structures in highly seismic areas, there should be many more opportunities to study their behavior.

When isolation and damping devices are not used, a sacrificial element will often limit the force and increase the damping in a structure. New RC beams and columns are provided with welded hoops and spirals that allow these members to form plastic hinges during large earthquakes. Existing concrete members are sometimes wrapped in steel casings, fiberglass, carbon fiber, and many other materials to increase their shear capacity and allow for the formation of a plastic hinge (Figure 2.89).

Steel structures have undergone similar improvements. Beginning in the late 1970s *eccentrically braced frames* (EBFs) were developed that provide greater stiffness and ductility during earthquakes. The EBF has a ductile link between the connections that is specially designed to act as an energy dissipater. This concept has been expanded to include a variety of different configurations (Figure 2.90).

The poor performance at the connections of moment-resisting frames during the Northridge and Kobe Earthquakes has resulted in a great deal of research and testing. There are now a variety of welded beam–column joints that ensure ductile behavior in the members rather than brittle fracture of the joint. One popular new connection is the "dog bone" (Figure 2.91). Testing has shown that plastic hinging will occur at the reduced section, protecting the connection.

There are a number of other devices and structural elements that improve structural response during large earthquakes. Restrainers, shear keys, catchers, and seat extenders are used to prevent bridge superstructures from falling from their supports. A variety of materials are wrapped around weak columns to increase their ductility. However, isolation and damping devices show the greatest potential

FIGURE 2.89 Steel shell being wrapped around a concrete bridge column to increase its ductility and shear strength.

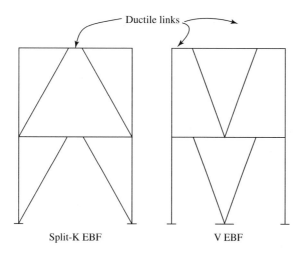

FIGURE 2.90 Two of the many configurations that have been developed for eccentrically braced connections.

to control the amount of damage on a structure, particularly when the goal is to keep the structure in service after a very large earthquake.

By ensuring that the supporting soil remains undamaged, by avoiding sites near active faults or other secondary hazards, and by the effective use of isolation and damping devices, most serious earthquake damage can be avoided. However, every structure should also be provided with abundant ductility and

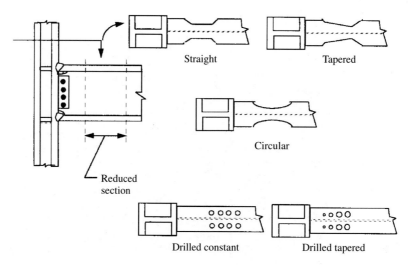

FIGURE 2.91 Examples of beams in moment-resisting frames that use a reduced flange section to prevent fracturing of the connection.

large seats in case an expectedly large earthquake were to occur. Moreover, all structures must be carefully designed to be relatively uniform and without eccentric loads.

References

[1] EERI, Scenario for a magnitude 7.0 earthquake on the Hayward fault, Earthquake Engineering Research Institute, HF-96, The Institute, Oakland, CA, 1996, 109 pages.

[2] Mualchin, L. and Jones, A.L., Peak acceleration from maximum credible earthquakes in California (rock and stiff-soil sites), DMG Open-file Report, 92-1, California Division of Mines and Geology, Sacramento, CA, 1992, 53 pages.

[3] National Research Council (U.S.) Committee on the Alaska Earthquake, The great Alaska earthquake of 1964, Publication (National Research Council [U.S.]) no. 1603, National Academy of Sciences, Washington, DC, 1968–1973, 8 v. in 10.

[4] Housner, G., et al., Competing against time, Report to the Governor George Deukmejian from The Governors' Board of Inquiry on the 1989 Loma Prieta earthquake, State of California Office of General Services, May 1990.

[5] EERI, The Mexico earthquake of September 19, 1985 — on the seismic response of the Valley of Mexico, *Earthquake Spectra*, 4, 3, August 1988, pages 569–589.

[6] USGS, The Loma Prieta, California Earthquake of October 17, 1989, United States Government Printing Office, 1998.

[7] NIST, The January 17, 1995 Hyogoken-Nanbu (Kobe) earthquake: performance of structures, lifelines, and fire protection systems, NIST Special Publication 901 (ICSSC TR18), U.S. National Institute of Standards and Technology, Gaithersburg, MD, July 1996, 538 pages.

[8] Japan Society of Civil Engineers (JSCE), Preliminary report on the Great Hanshin earthquake, January 17, 1995, Japan Society of Civil Engineers, 1995.

[9] EQE, Steel's performance in the Northridge earthquake, *EQE Int. Rev.*, Fall 1994, pages 1–6.

[10] Mitchell, J.K., Baxter, C.D.P., and Munson, T.C., Performance of improved ground during earthquakes, Soil Improvement for Earthquake Hazard Mitigation, American Society of Civil Engineers, New York, 1995, pages 1–36.

[11] Zayas, V.A. Low, S.S., and Mahin, S.A., The FPS earthquake resisting system: experimental report, UCB/EERC-87/01, Earthquake Engineering Research Center, University of California, Berkeley, CA, June 1987, 98 pages.

3

Seismic Design of Buildings

Ronald O. Hamburger
Simpson Gumpertz & Heger, Inc.,
San Francisco, CA

Charles Scawthorn
Department of Urban Management,
Kyoto University,
Kyoto, Japan

3.1 Introduction

Seismic design involves two distinct steps — determining (or estimating) the forces that will act on a structure and designing the structure so as to both resist these forces and keep deflections within prescribed limits. This chapter provides a basic explanation of the seismic design of buildings, by first discussing how earthquake forces are caused in buildings and the systems that have been developed to deal with these forces, then reviewing the most common types of buildings and their typical seismic performance, then discussing selected key aspects of seismic design, and, finally, discussing the force-determination aspect currently used in building codes.

3.2 Earthquakes — Their Cause and Effect

As discussed in Chapter 1, displacement of a fault results in the propagation of time-varying displacements throughout the earth — this vibratory ground motion is what is termed an *earthquake*. From a structural viewpoint, the essence of earthquakes is the dynamic displacement of the ground supporting a building, resulting in lateral and vertical forces on the building. In this chapter, emphasis will be on the lateral forces on a building due to shaking — vertical forces due to shaking are usually of somewhat less significance, but not always, and building codes and good practice require that vertical forces also be considered in the seismic design of buildings. While earthquakes also result in other effects on buildings,

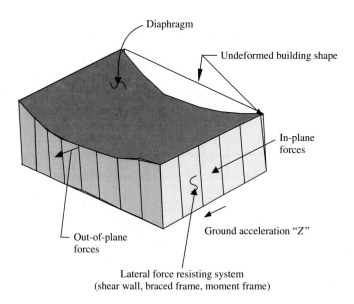

FIGURE 3.1 Effect of ground displacements on a building.

such as partial or complete ground failure, fire, and tsunami, design for these effects is more specialized and will not be covered here.

For a structure connected to the earth (and very few buildings are not connected to the earth through their base or foundation), as the base moves or displaces, the inertia and flexibility of the structure result in a time lag before the rest of the structure can displace in response to its base's displacement. The interaction of these displacements and the response of the structure result in time-varying displacements and strains within the structure, which the structure must be designed to sustain. For a structure responding to a moving base there is an equivalent system, in which the base is fixed and the structure is acted upon by forces (called inertia forces) that cause the same displacements as are occurring in the moving base system. In seismic design it is customary to visualize the structure as a fixed base system acted upon by inertia forces. Figure 3.1 shows such a system, in which the base has moved (due to the ground accelerations) and the roof is deflected. The roof's deflection is relative to its base, and is due to it and its supporting walls not moving or displacing as quickly as the base has displaced underneath it. This is exactly equivalent as if a force had been applied to the roof, which deflected the roof and its supporting walls the same amount, relative to its base.

3.3 Lateral Force Resisting Systems

The parts of the structure that connect the structure's mass to the ground and resist or otherwise accommodate these displacements or equivalent forces are termed the *lateral force resisting system* (LFRS). An LFRS is usually capable of resisting only forces that result from ground motions parallel to them. However, the combined action of LFRS along the width and length of a building can typically resist earthquake motion from any direction. LFRS differ from building to building because the type of system is controlled to some extent by the basic layout and structural elements of the building. Figure 3.2 illustrates the basic elements that may be used in LFRS, which consist of axial (tension and/or compression bracing) elements, shear (wall) elements, and/or bending resistant (frame) elements. Horizontal load distributing elements are termed diaphragms and are most often floor or roof slabs, but can be horizontally braced (i.e., truss) elements. Few buildings would use all of these LFRS elements, but most buildings would use one or more.

The earthquake resisting systems in modern steel buildings take many forms, Figure 3.3. Many types of bracing configurations have been used (diagonal, "X," "V," "K," etc.). Moment resisting steel frames

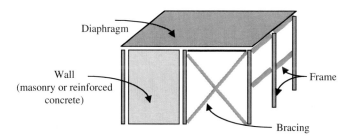

FIGURE 3.2 Types of elements comprising an LFRS.

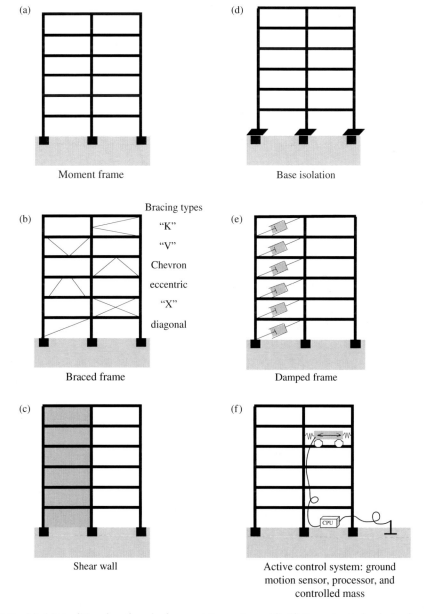

FIGURE 3.3 (a)–(c) Traditional earthquake force resisting systems. (d)–(f) Emerging technologies for earthquake force reducing systems.

are also capable of resisting lateral loads. In this type of construction, the connections between the beams and the columns are designed to resist the rotation of the column relative to the beam. Thus, the beam and the column work together and resist lateral movement by bending. This is contrary to the braced frame, where loads are resisted through tension and compression forces in the braces. Steel buildings are sometimes constructed with moment resistant frames in one direction and braced frames in the other, or with integral concrete or masonry shear walls.

In concrete and masonry structures, moment-resisting frames (MRFs) or *shear walls*[1] are used to provide lateral resistance. Ideally, these shear walls are continuous walls extending from the foundation to the roof of the building and can be exterior or interior walls. They are interconnected with the rest of the concrete frame and thus resist the motion of one floor relative to another. Shear walls can be constructed of cast-in-place reinforced concrete, precast concrete, reinforced brick, or reinforced hollow concrete block. Steel shear walls have been employed, but are not common.

3.3.1 Innovative Techniques

In about the last two decades, a number of innovative techniques have been developed and introduced to enable buildings and structures to better withstand earthquake ground motions. The general aim of these techniques, which are also shown in Figure 3.3, has been the *avoidance* of earthquake-induced forces, rather than their resistance. Innovative techniques can be divided into two broad categories, *passive control* (base isolation, energy dissipation) and *active control*, which are being increasingly applied to the design of new structures or to the retrofit of existing structures against wind, earthquakes, and other external loads (Soong and Costantinou 1994; Iemura and Pradono 2002; Yang et al. 2002).

3.3.1.1 Passive Control

The distinction between passive and active controls is that passive systems require no active intervention or energy source, while active systems typically monitor the structure and incoming ground motion and seek to actively control masses or forces in the structure (via moving weights, variable tension tendons, etc.) so as to develop a structural response (ideally) equal and opposite to the structural response due to the incoming ground motion. Recently developed semiactive control systems appear to combine the best features of both approaches, offering the reliability of passive devices and yet maintaining the versatility and adaptability of fully active systems. Magnetorheological (MR) dampers, for example, are new semiactive control devices that use MR fluids to create a controllable damper.

Passive control includes several categories: *base isolation* consists of softening of the shear capacity of a structure's connection with the ground, while maintaining vertical load-carrying capacity, so as to reduce the earthquake ground motion input to the structure. This has mostly been accomplished to date via the use of various types of rubber, lead–rubber, rubber–steel composite, or other types of bearings beneath columns. Key aspects of most of the base isolation systems developed to date are (a) economically limited to selected classes of structures (not too tall or short); (b) require additional foundation expense, including special treatment of incoming utility lines; and (c) require a certain amount of "rattlespace" around the structure to accommodate the additional displacements that the bearings will undergo. For new structures, these requirements are not especially onerous and a number of new structures in Japan and a few in other countries have been designed for base isolation. Most applications of this technology in the United States, however, have been for the retrofit of existing (usually historic) structures, where the technology permitted increased seismic capacity without major modification to the architectural features. The technique has been applied to a number of highway bridges and some industrial structures in the United States.

---

[1]Termed shear walls because the depth to width ratio is so large that deformation is primarily due to shear rather than bending.

3.3.1.1.1 Supplemental Damping

If damping can be significantly increased, then structural responses (and therefore forces and displacements) are greatly reduced (Hanson et al. 1993). Supplemental damping systems include *friction* systems (e.g., Sumitomo, pall, and friction-slip) based on Coulomb friction, *self-centering* friction resistance that is proportional to displacement (e.g., Fluor–Daniel energy dissipating restraint), or various *energy dissipation* mechanisms: ADAS (added damping and stiffness) elements, which utilize the yielding of mild-steel X-plates; viscoelastic shear dampers using a 3M acrylic copolymer as the dissipative element; or nickel–titanium alloy shape-memory devices that take advantage of reversible, stress-induced phase changes in the alloy to dissipate energy (Aiken et al. 1993). These systems are generally still in the developmental stage although there are no special obstacles for the implementation of the ADAS system, which has seen one application to date in the United States (Perry et al. 1993) and more in other countries.

3.3.1.2 Active Control

Active control depends on actively modifying a structure's mass, stiffness, or geometric properties during its dynamic response in such a manner so as to counteract and reduce excessive displacements (Iemura and Pradono 2002). Tuned mass dampers, reliance on liquid sloshing (Lou et al. 1994), and active tensioning of tendons are methods currently under investigation. Most methods of active control are real time, relying on measurement of structural response, rapid structural computation, and fast-acting energy sources. A number of issues of reliability remain to be resolved (Spencer et al. 1994).

3.4 Types of Buildings and Typical Earthquake Performance

There are many different types of buildings, with varying kinds of earthquake performance and seismic design needs. This section discusses general earthquake performance of buildings, with the emphasis more toward those buildings typically built in the western United States. Specific aspects of structural analysis and design of buildings, other structures, steel, concrete, wood, masonry, and other topics are discussed in other chapters.

In buildings, earthquake performance can be divided into two categories: structural and non-structural, both of which when unsatisfactory can be hazardous to building occupants — when *damage* occurs. Structural damage means degradation of the building's structural support systems (i.e., vertical and lateral force resisting systems), such as the building frames and walls. Nonstructural damage refers to any damage that does not affect the integrity of the structural support system. Examples of nonstructural damage are a chimney collapsing, windows breaking, ceilings falling, piping damage, and disruption of pumps, control panels, telecommunications equipment, etc. Nonstructural damage can still be life threatening and costly. The type of damage to be expected is a complex issue that depends on the structural type and age of the building, its configuration, construction materials, the site conditions, the proximity of the building to neighboring buildings, and the type of nonstructural elements.

The typical earthquake performances of different types of common building structural systems are described in this section to provide insights into seismic design for buildings.

3.4.1 Wood Frame

Wood-frame structures tend to be mostly low rise (one to three stories, occasionally four stories). Vertical framing may be of several types, for example, stud wall, braced post and beam, or timber pole:

- Stud walls are typically constructed of 2 in. by 4 in. wood members vertically set about 16 in. apart — multiple story buildings may have 2×6 or larger studs. These walls are braced by plywood sheathing or by diagonals made of wood or steel. Most detached single and low-rise multiple family residences in the United States are of stud wall wood frame construction, Figure 3.4.

Roof/floor span systems
1. Wood joist and rafter
2. Diagonal sheathing
3. Straight sheathing

Wall systems
4. Stud wall (platform or balloon frame)
5. Horizontal siding

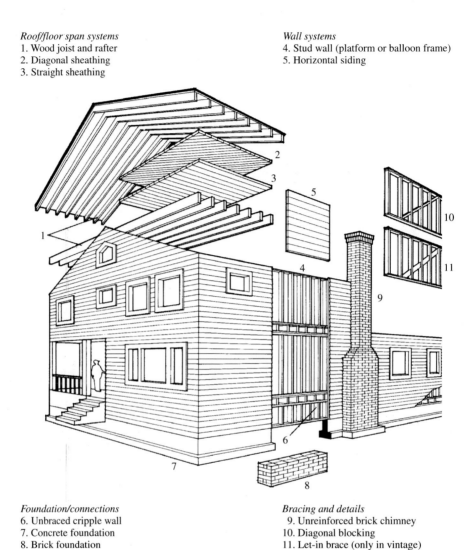

Foundation/connections
6. Unbraced cripple wall
7. Concrete foundation
8. Brick foundation

Bracing and details
9. Unreinforced brick chimney
10. Diagonal blocking
11. Let-in brace (only in vintage)

FIGURE 3.4 Schematic of wood light-frame construction. (From Federal Emergency Management Agency. 1988. *Rapid Visual Screening of Buildings for Potential Seismic Hazards: A Handbook,* FEMA 154, FEMA, Washington, DC.)

- Post and beam construction is not very common in the United States, although it is the basis of the traditional housing in other countries (e.g., Europe, Japan); in the United States, it is found mostly in older housing and larger buildings (i.e., warehouses, mills, churches, and theaters). This type of construction consists of larger rectangular (6 in. by 6 in. and larger) or sometimes round wood columns framed together with large wood beams or trusses.
- Timber pole buildings are a less common form of construction found mostly in suburban/rural areas. Generally adequate seismically when first built, they are more often subject to wood deterioration due to the exposure of the columns, particularly near the ground surface. Together with an often-found "soft story" in this building type, this deterioration may contribute to unsatisfactory seismic performance.

In wood frame stud-wall buildings, the resistance to lateral loads is typically provided by (a) for older buildings, especially houses, wood diagonal "let-in" bracing and (b) for newer (primarily post-World War II) buildings, plywood siding "shear walls." Without the extra strength provided by the bracing or plywood, walls would distort excessively or "rack," resulting in broken windows, stuck doors, cracked plaster, and, in extreme cases, collapse.

Stud-wall buildings have performed very well in past U.S. earthquakes for ground motions of about $0.5g$ or less, due to inherent qualities of the structural system and because they are lightweight and low rise. Cracking in plaster and stucco may occur and these act to degrade the strength of the building to some extent (i.e., the plaster and stucco may in fact form part of the LFRS, sometimes by design) — this is usually classified as nonstructural damage but, in fact, dissipates a lot of the earthquake-induced energy. However, the most common type of structural damage in older wood-frame buildings results from a lack of connection between the superstructure and the foundation — the so-called "cripple wall" construction. This kind of construction is common in the milder climes of the west, where full basements are not required, and consists of an air space (typically 2–3 ft) left under the house — the short stud walls under the first floor (termed by carpenters a cripple wall because of their less than full height) were usually built without bracing so that their is no adequate LFRS for this short height. Plywood sheathing nailed to the cripple studs may have been used to strengthen the cripple walls. Additionally, the mud sill in these older (typically pre-World War II) housing may not be bolted to the foundation. As a result, houses can slide off their foundations when not properly bolted to the foundation, resulting in major damage to the building as well as to plumbing and electrical connections. Overturning of the entire structure is usually not a problem because of the low-rise geometry. In many municipalities, modern codes require wood structures to be bolted to their foundations. However, the year that this practice was adopted will differ from community to community and should be checked.

Garages often have a very large door opening in one wall with little or no bracing. This wall has almost no resistance to lateral forces, which is a problem if a heavy load such as a second story sits on top of the garage (the so-called house over garage, or HOGs). Homes built over garages have sustained significant amounts of damage in past earthquakes, with many collapses. Therefore, the HOG configuration, which is found commonly in low-rise apartment complexes and some newer suburban detached dwellings, should be examined more carefully and perhaps strengthened.

Unreinforced masonry (URM) chimneys also present a life-safety problem. They are often inadequately tied to the building and therefore fall when strongly shaken. On the other hand, chimneys of reinforced masonry generally perform well.

Some wood-frame structures, especially older buildings in the eastern United States, have masonry veneers that may represent another hazard. The veneer usually consists of one wythe of brick (a wythe is a term denoting the width of one brick) attached to the stud wall. In older buildings, the veneer is either insufficiently attached or has poor quality mortar, which often results in peeling off of the veneer during moderate and large earthquakes.

Post and beam buildings tend to perform well in earthquakes if adequately braced. However, walls often do not have sufficient bracing to resist horizontal motion and thus they may deform excessively.

The 1994 M_W 6.7 Northridge earthquake was the largest earthquake to occur directly within an urbanized area since the 1971 San Fernando earthquake — ground motions were as high as $0.9g$ and substantial numbers of modern wood-frame dwellings sustained significant damage, including major cracking of veneers, gypsum board walls, and splitting of wood wall studs. It may be inferred from this, as well as the performance observed in the more sparsely populated epicentral regions of the 1989 M_W 7.1 Loma Prieta Earthquake, that U.S. single family dwelling design begins to sustain substantial nonstructural and structural damage for peak ground acceleration in excess of about $0.5g$.

3.4.2 Steel-Frame Buildings

Steel-frame buildings generally may be classified as MRFs, braced frames, or mixed construction (e.g., steel frame for vertical forces and reinforced concrete shear wall for the LFRS) based on their LFRSs. In concentric braced frames the lateral forces or loads are resisted by the tensile and compressive strength of the bracing, which can assume a number of different configurations including diagonal, "V," inverted "V" also termed chevron, "K," etc. A recent development in seismic bracing is the eccentric brace frame. Here, the bracing is slightly offset from the main beam to column connection, and the short section of the beam is expected to deform significantly under major seismic forces and thereby dissipate a considerable portion of the energy. MRFs resist lateral loads and deformations by the bending stiffness of the beams and columns (there is no bracing), Figure 3.5.

Steel-frame buildings have tended to perform satisfactory in earthquakes with ground motions less than about 0.5g because of their strength, flexibility, and lightness. Collapse in earthquakes has been very rare, although steel-frame buildings did collapse, for example, in the 1985 Mexico City Earthquake. More recently, following the 1994 M_W 6.7 Northridge Earthquake, a number of MRFs were found to have sustained serious cracking in the beam column connection; see Figure 3.6, which shows one of a number of different types of cracking that were found following the Northridge Earthquake. The cracking typically initiated at the lower beam flange location and propagated upward into the shear panel. Similar cracking was also observed following the 1995 M_W 6.9 Hanshin (Kobe) Earthquake, which experienced similar levels of ground motion as Northridge. More worrisome is that, as of this writing, some steel buildings in the San Francisco Bay Area have been found to have similar cracking, presumably as a result of the 1989 M_W 7.1 Loma Prieta Earthquake. As a result, there is an ongoing effort by a consortium of research organizations, termed SAC (funded by the Federal Emergency Management Agency) to better understand and develop solutions for this problem.

Light-gage steel buildings are used for agricultural structures, industrial factories, and warehouses. They are typically one story in height, sometimes without interior columns, and often enclose a large floor area, Figure 3.7. Construction is typically of steel frames spanning the short dimension of the building and resisting lateral forces as moment frames. Forces in the long direction are usually resisted by diagonal steel rod bracing. These buildings are usually clad with lightweight metal or asbestos reinforced concrete siding, often corrugated. Because these buildings are low rise, lightweight, and constructed of steel members, they usually perform relatively well in earthquakes. Collapses do not usually occur. Some typical problems are (a) insufficient capacity of tension braces can lead to their elongation and, in turn, building damage and (b) inadequate connection to the foundation can allow the building columns to slide.

3.4.3 Concrete Buildings

Several construction subtypes fall under this category: (a) MRFs (nonductile or ductile); (b) shear wall structures; and (c) precast, including tilt-up structures. The most prevalent of these is nonductile reinforced concrete frame structures with or without infill walls built in the United States between about 1920 and (in the western United States) 1972. In many other portions of the United States this type of construction continues to the present. This group includes large multistory commercial, institutional, and residential buildings constructed using flat slab frames, waffle slab frames, and the standard girder-column-type frames. These structures generally are more massive than steel frame buildings, are underreinforced (i.e., have insufficient reinforcing steel embedded in the concrete), and display low ductility. Some typical problems are (a) large tie spacings in columns can lead to a lack of concrete confinement and/or shear failure; (b) placement of inadequate rebar splices at the same location can lead to column failure; (c) insufficient shear strength in columns can lead to shear failure prior to the development of moment hinge capacity; (d) insufficient shear tie anchorage can prevent the column from developing its full shear capacity; (e) lack of continuous beam reinforcement can result in hinge formation during load reversal; (f) inadequate reinforcing of beam–column joints or location of beam bar splices at columns can lead to failures; and (g) the relatively low stiffness of the frame can lead to substantial nonstructural damage.

Ductile reinforced concrete frames where special reinforcing details are required in order to furnish satisfactory load-carrying performance under large deflections (termed ductility) have usually only been required in the highly seismic portions of the United States since the mid-1970s. ACI-318 (1995)

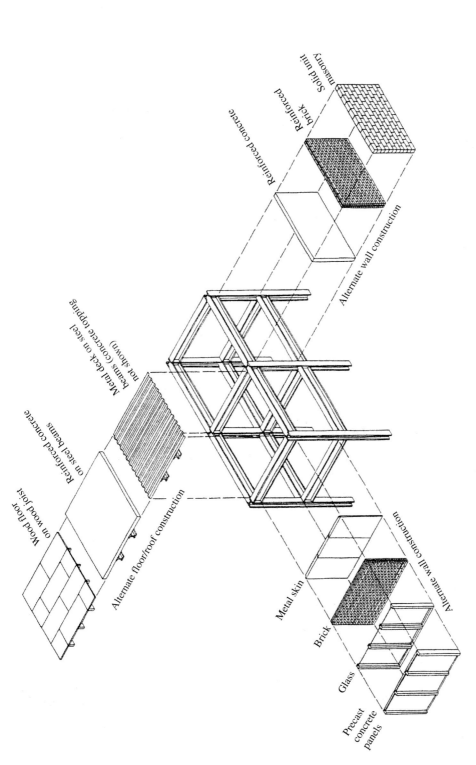

FIGURE 3.5 Steel moment resisting frame construction. (From Federal Emergency Management Agency. 1988. *Rapid Visual Screening of Buildings for Potential Seismic Hazards: A Handbook*, FEMA 154, FEMA, Washington, DC.)

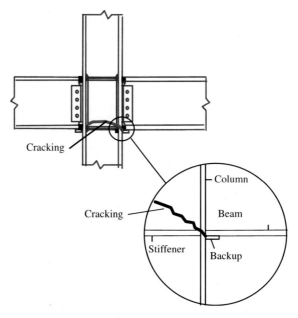

FIGURE 3.6   Example steel moment–frame connection cracking.

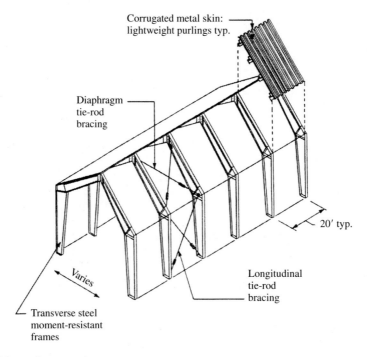

FIGURE 3.7   Light steel construction.

provides comprehensive treatment for the *ductile detailing*, which involves a number of special requirements including close spacing of lateral reinforcement in order to attain *confinement* of a concrete core, appropriate relative dimensioning of beams and columns, 135° hooks on lateral reinforcement, hooks on main beam reinforcement within the column, etc.

Concrete shear wall buildings consist of a concrete box or frame structural system with walls constituting the main LFRS, Figure 3.8. The entire structure, along with the usual concrete diaphragm, is typically cast in place. Shear walls in buildings can be located along the perimeter, as interior walls, or around the service or elevator core. This building type generally tends to perform better than concrete frame buildings. They are heavier than steel-frame buildings but they are also rigid due to the shear walls. Some types of damage commonly observed in taller buildings are caused by vertical discontinuities, pounding, and/or irregular configuration. Other damages specific to this building type are (a) shear cracking and distress can occur around openings in concrete shear walls during large seismic events; (b) shear failure can occur at wall construction joints usually at a load level below the expected capacity; and (c) bending failures can result from insufficient chord steel lap lengths.

Tilt-up buildings are a common type of construction in the western United States and consist of concrete wall panels cast on the ground and then tilted upward into their final positions. More recently, wall panels are fabricated off-site and trucked in. The wall panels are welded together at embedments or held in place by cast-in-place columns or steel columns, depending on the region. The floor and roof beams are often glue-laminated wood or steel open webbed joists that are attached to the tilt-up wall panels; these panels may be load bearing or nonload bearing, depending on the region. These buildings tend to be low-rise industrial or office buildings. Before 1973 in the western United States, many tilt-up buildings did not have sufficiently strong connections or anchors between the walls and the roof and floor diaphragms. During an earthquake, weak anchors pull out of the walls, causing the floors or roofs

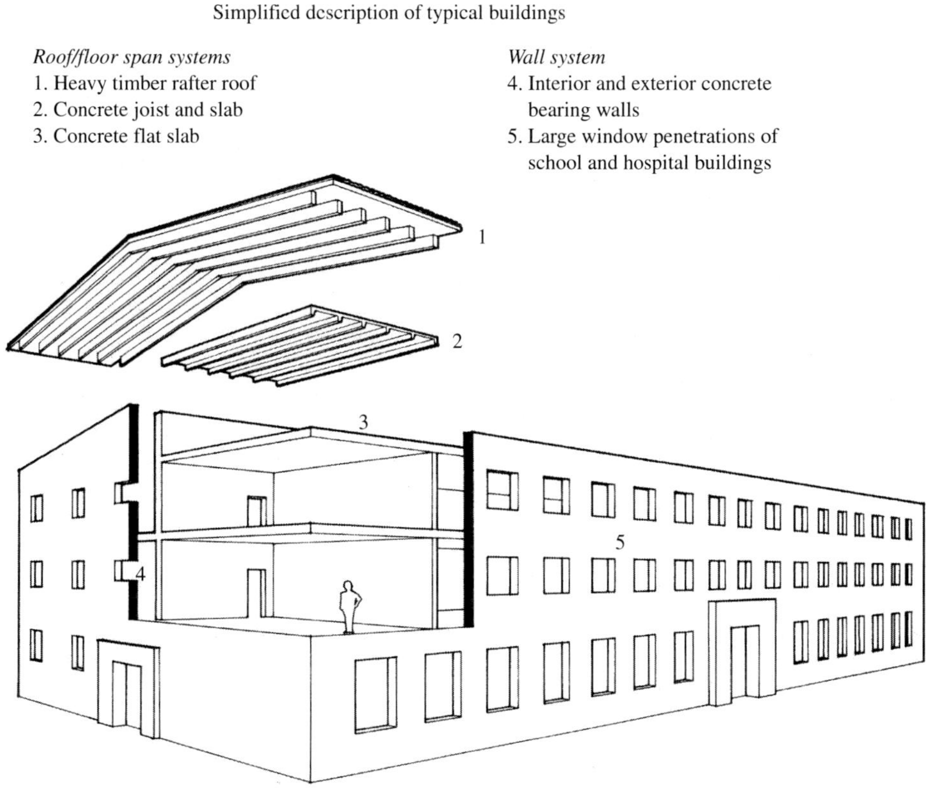

Simplified description of typical buildings

Roof/floor span systems
1. Heavy timber rafter roof
2. Concrete joist and slab
3. Concrete flat slab

Wall system
4. Interior and exterior concrete bearing walls
5. Large window penetrations of school and hospital buildings

FIGURE 3.8   Reinforced concrete shear wall construction.

to collapse. The connections between concrete panels are also vulnerable to failure. Without these, the building loses much of its lateral force resisting capacity. For these reasons, many tilt-up buildings were damaged in the 1971 San Fernando Earthquake. Since 1973, tilt-up construction practices have changed in California and other high-seismicity regions, requiring positive wall–diaphragm connection and prohibiting cross-grain bending in wall ledgers. (Such requirements may not have yet been made in other regions of the country.) However, a large number of these older, pre-1970s vintage tilt-up buildings still exist and have not been retrofitted to correct this wall-anchor defect. These buildings are a prime source of seismic hazards. In areas of low or moderate seismicity, inadequate wall anchor details continue to be employed. Damage to tilt-up buildings was observed again in the 1994 M_W 6.7 Northridge earthquake, where the primary problems were poor wall anchorage into the concrete and excessive forces due to flexible roof diaphragms amplifying ground motion to a greater extent than anticipated in the code.

Precast concrete frame construction, first developed in the 1930s, was not widely used until the 1960s. The precast frame is essentially a post and beam system in concrete where columns, beams, and slabs are prefabricated and assembled on site, Figure 3.9. Various types of members are used: vertical load-carrying elements may be Ts, cross-shapes, or arches and are often more than one story in height. Beams are often Ts and double Ts or rectangular sections. Prestressing of the members, including pretensioning and posttensioning, is often employed. The LFRS is often concrete cast-in-place shear walls. The earthquake performance of this structural type varies greatly and is sometimes poor. This type of building can perform well if the details used to connect the structural elements have sufficient strength and ductility (toughness). Because structures of this type often employ cast-in-place concrete shear walls for lateral load resistance, they experience the same types of damage as other shear wall building types. Some of the problem areas specific to precast frames are (a) poorly designed connections between

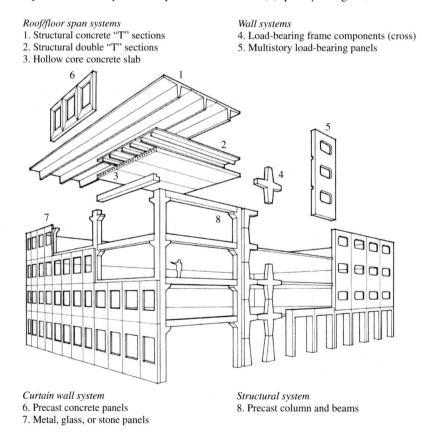

Roof/floor span systems
1. Structural concrete "T" sections
2. Structural double "T" sections
3. Hollow core concrete slab

Wall systems
4. Load-bearing frame components (cross)
5. Multistory load-bearing panels

Curtain wall system
6. Precast concrete panels
7. Metal, glass, or stone panels

Structural system
8. Precast column and beams

FIGURE 3.9   Precast concrete construction.

prefabricated elements can fail; (b) accumulated stresses can result due to shrinkage and creep and due to stresses incurred in transportation; (c) loss of vertical support can occur due to inadequate bearing area and/or insufficient connection between floor elements and columns; and (d) corrosion of metal connectors between prefabricated elements can occur. A number of precast parking garages failed in the 1994 M_W 6.7 Northridge Earthquake, including a large structure at the Cal State Northridge campus that sustained a progressive failure. This structure had a perimeter precast MRF and interior non-ductile columns — the MRF sustained large but tolerable deflections; however, interior nonductile columns failed under these deflections, resulting in an interior collapse, which then pulled the exterior MRFs over.

3.4.4 Masonry Buildings

Reinforced masonry buildings are mostly low-rise perimeter bearing wall structures, often with wood diaphragms although precast concrete is sometimes used. Floor and roof assemblies usually consist of timber joists and beams, glue-laminated beams, or light steel joists. The bearing walls consist of grouted and reinforced hollow or solid masonry units. Interior supports, if any, are often wood or steel columns, wood stud frames, or masonry walls. Generally, they are less than five stories in height although many mid-rise masonry buildings exist. Reinforced masonry buildings can perform well in moderate earthquakes if they are adequately reinforced and grouted and if sufficient diaphragm anchorage exists.

Most URM bearing wall structures in the western United States were built before 1934, although this construction type was permitted in some jurisdictions having moderate or high seismicity until the late 1940s or early 1950s (in low-seismicity jurisdictions URM may still be a common type of construction, even today). These buildings usually range from one to six stories in height and typically construction varies according to the type of use, although wood floor and roof diaphragms are common. Smaller commercial and residential buildings usually have light wood floor/roof joists supported on the typical perimeter URM wall and interior wood load-bearing partitions. Larger buildings, such as industrial warehouses, have heavier floors and interior columns, usually of wood. The bearing walls of these industrial buildings tend to be thick, often as much as 24 in. or more at the base. Wall thicknesses of residential buildings range from 9 in. at upper floors to 18 in. at lower floors. URM structures are recognized as perhaps the most hazardous structural type. They have been observed to fail in many modes during past earthquakes. Typical problems are

1. *Insufficient anchorage.* Because the walls, parapets, and cornices are not positively anchored to the floors, they tend to fall out. The collapse of bearing walls can lead to major building collapses. Some of these buildings have anchors as a part of the original construction or as a retrofit. These older anchors exhibit questionable performance.
2. *Excessive diaphragm deflection.* Because most of the floor diaphragms are constructed of wood sheathing, they are very flexible and permit large out-of-plane deflection at the wall transverse to the direction of the force. The large drift, occurring at the roof line, can cause the masonry wall to collapse under its own weight.
3. *Low shear resistance.* The mortar used in these older buildings is often made of lime and sand, with little or no cement, and has very little shear strength. The bearing walls will be heavily damaged and collapse under large loads.
4. *Wall slenderness.* Some of these buildings have tall story heights and thin walls. This condition, especially in nonload-bearing walls, will result in buckling out-of-plane under severe lateral load. Failure of a nonload-bearing wall represents a falling hazard, whereas the collapse of a load-bearing wall will lead to partial or total collapse of the structure.

3.4.5 Configuration, Irregularities, and Pounding

Certain problems in earthquake performance are common to many building types and include issues of configuration, irregularities, and pounding.

3.4.5.1 Configuration and Irregularities

Configuration, or the general vertical and/or horizontal shape of buildings, is an important factor in earthquake performance and damage. Buildings that have simple, regular, symmetric configurations generally display the best performance in earthquakes. The reasons for this are (a) nonsymmetric buildings tend to have twist (i.e., have significant torsional modes) in addition to shaking laterally and (b) the various "wings" of a building tend to act independently, resulting in differential movements, cracking, and other damage. Rotational motion introduces additional damage, especially at re-entrant or "internal" corners of the building. The term "configuration" also refers to the geometry of lateral load resisting systems as well as the geometry of the building. Asymmetry can exist in the placement of bracing systems, shear walls, or MRFs that are used to provide earthquake resistance in a building. This type of asymmetry, of the LFRS, can result in twisting or differential motion, with the same consequences as asymmetry in the building plan. An important aspect of configuration is *soft story,* which is a story of a building signifiantly less stiff than adjacent stories (i.e., a story in which the lateral stiffness is 70% or less than that in the story above or less than 80% of the average stiffness of the three stories above; BSSC 2001). Soft stories often (but not always) occur on the ground floor, where commercial or other reasons require a greater story height, and large windows or openings for ingress or commercial display (e.g., the building might have masonry curtain walls for the full height, except at the ground floor, where these are replaced with large windows for a store's display). Due to inadequate stiffness, a disproportionate amount of the entire building's drift is concentrated at the soft story, resulting in nonstructural and potential structural damage. Many older buildings with soft stories but built prior to recognition of this aspect collapse due to excessive ductility demands at the soft story.

The National Earthquake Hazard Reduction Program (NEHRP) provisions for the design of new buildings (BSSC 2001) have defined when a building's configuration is "irregular," and provided required strength increase factors or other approaches to deal with these irregularities. The NEHRP definitions for plan and vertical irregularities are illustrated in Table 3.1 and Table 3.2.

3.4.5.2 Pounding

Pounding is the collision of adjacent buildings during an earthquake due to insufficient lateral clearance. Such collision can induce very high and unforeseen accelerations and story shears in the overall structure. Additionally, if adjacent buildings have varying story heights, a relatively rigid floor or roof diaphragm may impact adjacent buildings at or near mid-column height, causing bending or shear failure in the columns and subsequent story collapse. Under earthquake lateral loading, buildings deflect significantly — these deflections or drift are limited by code — and adjacent buildings must be separated by a seismic gap equal to the sum of their actual calculated drifts (i.e., ideally, each building set back from its property line by the drift). Pounding has been the cause of a number of mid-rise building collapses, most notably in the 1985 Mexico City Earthquake.

3.5 2000 NEHRP Recommended Provisions

3.5.1 Overview

The *2000 NEHRP Recommended Provisions for Seismic Regulation for Buildings and Other Structures* (NEHRP Provisions) represents the current state of the art in prescriptive, as opposed to performance-based, provisions for seismic-resistant design. Its provisions form the basis for earthquake design specifications contained in the 2001 edition of *ASCE-7, Minimum Design Loads for Buildings and other Structures,* either through reference or direct incorporation, the seismic regulations in the 2003 edition of the *International Building Code,* and also the 2002 edition of the *NFPA 5000 Building Code* (NFPA, n.d.). As such, it will form the basis for most earthquake-resistant design in the United States, as well as other nations that base their codes on U.S. practices, throughout much of the first decade of the twenty-first century.

TABLE 3.1 NEHRP 2000 Plan Structural Irregularities

1: Torsional irregularity — when diaphragms are not flexible

Torisonal irregularity exists (1a) when the maximum story drift, computed including accidental torsion, at one end of the structure transverse to an axis is more than 1.2 times the average of the story drift at the two ends of the structure. (1b) Extreme torisonal irregularity exists when ratio >1.4

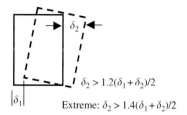

$\delta_2 > 1.2(\delta_1 + \delta_2)/2$

Extreme: $\delta_2 > 1.4(\delta_1 + \delta_2)/2$

2: Re-entrant corners

Plan configurations of a structure and its lateral-force-resisting system contain re-entrant corners where both projections of the structure beyond a re-entrant corner are greater than 15% of the plan dimension of the structure in the given direction

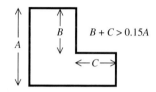

$B + C > 0.15A$

3: Diaphragam discontinuity

Diaphragms with abrupt discontinuities or variations in stiffness including those having cutout or open areas greater than 50% of the gross enclosed diaphragm area or changes in effective diaphragm stiffness of more than 50% from one story to the next

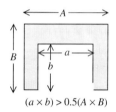

$(a \times b) > 0.5(A \times B)$

4: Out-of-plane offsets

Discontinuities in a lateral-force-resistance path such as out-of-plane offsets of the vertical elements

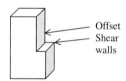

Offset
Shear walls

5: Nonparallel systems

The vertical lateral-force-resisting elements are not parallel to or symmetric about the major orthogonal axes of the lateral-force-resisting system

The *NEHRP Provisions* assume significant amounts of nonlinear behavior will occur under design level events. The extent of nonlinear behavior that may occur is dependent on the structural systems employed in resisting earthquake forces, the configuration of these systems, and the extent that the structural systems are detailed for ductile behavior under large cyclic inelastic deformation. The *NEHRP Provisions* may therefore be thought to consist of two component parts:

- One part relates to specification of the required design strength and stiffness of the structural system.
- The second part relates to issues of structural detailing.

TABLE 3.2   NEHRP 2000 Vertical Structural Irregularities

1: Stiffness irregularity — soft story

(1a) A soft *story* is one in which the lateral stiffness is less than 70% of that in the *story* above or less than 80% of the average stiffness of the three stories above. (1b) An extreme soft *story* is for ratios of less than 60% or less than 70%, respectively

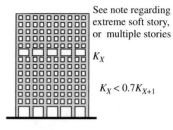

See note regarding extreme soft story, or multiple stories

K_X

$K_X < 0.7 K_{X+1}$

2: Weight (mass) irregularity

Mass irregularity exists where the effective mass of any *story* is more than 150% of the effective mass of an adjacent *story*. A roof that is lighter than the floor below need not be considered

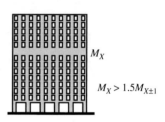

M_X

$M_X > 1.5 M_{X\pm1}$

3: Vertical geometric irregularity

Vertical geometric irregularity exists where the horizontal dimension of the lateral-force-resisting system in any *story* is more than 130% of that in an adjacent *story*

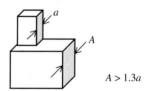

a

A

$A > 1.3a$

4: In-plane discontinuity in vertical lateral-force-resisting elements

An in-plane offset of the lateral-force-resisting elements greater than the length of those elements or a reduction in stiffness of the resisting elements in the story below

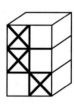

5: Discontinuity in capacity — weak story

A weak *story* is one in which the *story* lateral *strength* is less than 80% of that in the *story* above. The *story strength* is the total *strength* of all seismic-resisting elements sharing the *story* shear for the direction under consideration

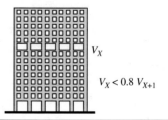

V_X

$V_X < 0.8\, V_{X+1}$

For this second part, the *NEHRP Provisions* adopt, with modification, design standards and specifications developed by industry groups such as the American Concrete Institute or the American Institute of Steel Construction. This second part of the *NEHRP Provisions* is not discussed in this chapter, but is covered in detail in each of the following chapters, which treat the individual structural materials.

Instead, this article focuses primarily on the manner in which the *NEHRP Provisions* regulate the required strength and stiffness of structures.

3.5.2 Performance Intent and Objectives

The *NEHRP Provisions* are intended to provide a tiered series of performance capabilities for structures, depending on their intended occupancy and use. Under the *NEHRP Provisions*, each structure must be assigned to a seismic use group (SUG). Three SUGs are defined and are respectively labeled I, II, and III:

- SUG-I encompasses most ordinary occupancy buildings including typical commercial, residential, and industrial structures. For these facilities the basic intent of the *NEHRP Provisions*, just as with earlier codes, is to provide a low probability of earthquake-induced life safety endangerment.
- SUG-II includes facilities that house large numbers of persons, persons who are mobility impaired, or large quantities of materials that if released could pose substantial hazards to the surrounding community. Examples of such facilities include large assembly facilities, housing several thousand persons, daycare centers, and manufacturing facilities containing large quantities of toxic or explosive materials. The performance intent for these facilities is to provide a lower probability of life endangerment, relative to SUG-I structures, and a low probability of damage that would result in release of stored materials.
- SUG-III includes those facilities such as hospitals and emergency operations and communications centers deemed essential to disaster response and recovery operations. The basic performance intent of the *NEHRP Provisions* with regard to these structures is to provide a low probability of earthquake-induced loss of functionality and operability.

In reality, the probability of damage resulting in life endangerment, release of hazardous materials, or loss of function should be calculated using structural reliability methods as the total probability of such damage over a period of time (Ravindra 1994). Mathematically, this is equal to the integral, over all possible levels of ground motion intensity, of the conditional probability of excessive damage given that a ground motion intensity is experienced and the probability that such ground motion intensity will be experienced in the desired period of time. Although such an approach would be mathematically and conceptually correct, it is currently regarded as too complex for practical application in the design office.

Instead, the *NEHRP Provisions* design for desired limiting levels of nonlinear behavior for a single design earthquake intensity level, termed *maximum considered earthquake* (MCE) ground shaking. In most regions of the United States, the MCE is defined as that intensity of ground shaking having a 2% probability of exceedance in 50 years. In certain regions, proximate to major active faults, this probabilistic definition of MCE motion is limited by a conservative deterministic estimate of the ground motion intensity anticipated to result from an earthquake of characteristic magnitude on these faults. The MCE is thought to represent the most severe level of shaking ever likely to be experienced by a structure, though it is recognized that there is some limited possibility of more severe motion occurring.

Structures categorized as SUG-I are designed with the expectation that MCE shaking would result in severe damage to both structural and nonstructural elements, with damage perhaps being so severe that following the earthquake the structure would be on the verge of collapse. This damage state has come to be termed *collapse prevention*, because the structure is thought to be at a state of incipient but not actual collapse. Theoretically, SUG-I structures behaving in this manner would be total or near total financial losses, in the event that MCE shaking was experienced. To the extent that shaking experienced by the structure exceeds the MCE level, the structure could actually experience partial or total collapse.

SUG-III structures are designed with the intent that when subjected to MCE shaking they would experience both structural and nonstructural damages; however, the structures would retain significant residual structural resistance or margin against collapse. It is anticipated that when experiencing MCE shaking such structures may be damaged to an extent that they would no longer be suitable for occupancy, until repair work had been instituted, but that repair would be technically and economically feasible. This superior performance relative to SUG-I structures is accomplished through specification that SUG-III structures be designed with 50% greater strength and more stiffness than their SUG-I counterparts. SUG-II structures are designed for performance intermediate to that for SUG-I and SUG-III with strengths and stiffness that are 25% greater than those required for SUG-I structures.

3.5.3 Seismic Hazard Maps and Ground Motion Parameters

The *NEHRP Provisions* incorporate a series of national seismic hazard maps for the United States and territories, developed by the United States Geologic Survey (USGS), specifically for this purpose (available at http://geohazards.cr.usfs.gov/eq/index.html). Two sets of maps are presented. One set presents contours of MCE, 5% damped, elastic spectral response acceleration at a period of 0.2 s, termed S_S. The second set presents contours of MCE, 5% damped, elastic spectral response acceleration at a period of 1.0 s, termed S_1. In both cases, the spectral response acceleration values are representative of sites with subsurface conditions bordering between firm soil or soft rock. Contours are presented in increments of $0.02g$ in areas of low seismicity and $0.05g$ in areas of high seismicity. By locating a site on the maps and interpolating between the values presented for contours adjacent to the site, it is possible to rapidly estimate the MCE level shaking parameters for the site, given that it has a soft rock or firm soil profile Figure 3.10 shows, for a portion of the western United States, contours of the 0.2 s spectral acceleration with a 90% probability of not being exceeded in 50 years. As indicated in the figure, in zones of high seismicity these contours are quite closely spaced, making use of the maps difficult. Therefore, the USGS has furnished software, available both over the internet (at the URL indicated above) and on a CD-ROM, which permits determination of the MCE spectral response acceleration parameters based on longitude and latitude.

Since many sites are located neither on soft rock nor on firm soil sites, it is necessary to correct the mapped values of spectral response acceleration to account for site amplification and de-amplification effects. To facilitate this process, a site is categorized into one of six site class groups, labeled A through F. Table 3.3 summarizes the various site class categories.

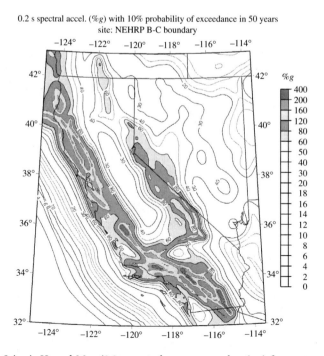

FIGURE 3.10   MCE Seismic Hazard Map (0.2 s spectral response acceleration) for western United States.

TABLE 3.3 Site Categories

Site class	Description	Shear wave velocity, $\bar{v}_s$	Penetration resistance, $\bar{N}$	Unconfined shear strength, $\bar{s}_u$
A	Hard rock	>5000 ft/s		
B	Rock	2500 ft/s $< \bar{v}_s \le$ 5000 ft/s		
C	Very firm soil or soft rock	1200 ft/s $< \bar{v}_s \le$ 2500 ft/s	>50	>2000 psf
D	Stiff soil	600 ft/s $< \bar{v}_s \le$ 1200 ft/s	15–50	1000–2000 psf
E	Soil	$\bar{v}_s <$ 600 ft/s	<15	<1000 psf
F	Special soils	Soils requiring site-specific evaluations: 1. Soils vulnerable to potential failure or collapse under seismic loading such as liquefiable soils, quick and highly sensitive clays, collapsible weakly cemented soils 2. Peats and/or highly organic clays ($H >$ 10 ft of peat and/or highly organic clay where $H =$ thickness of soil) 3. Very high plasticity clays ($H >$ 25 ft [8 m] with PI $>$ 75) 4. Very thick soft/medium stiff clays ($H >$ 120 ft [36 m])		

Note: $\bar{v}_s$, $\bar{N}$, $\bar{s}_u$ represent the average value of the parameter over the top 30 m (100 ft) of soil.

TABLE 3.4 Coefficient F_a as a Function of Site Class and Mapped Spectral Response Acceleration

Site class	Mapped maximum considered earthquake spectral response acceleration at short periods				
	$S_S = 0.25$	$S_S = 0.50$	$S_S = 0.75$	$S_S = 1.00$	$S_S = 1.25$
A	0.8	0.8	0.8	0.8	0.8
B	1.0	1.0	1.0	1.0	1.0
C	1.2	1.2	1.1	1.0	1.0
D	1.6	1.4	1.2	1.1	1.0
E	2.5	1.7	1.2	0.9	a
F	a	a	a	a	a

Note: "a" indicates site-specific evaluation required.

Once a site has been categorized within a site class, a series of coefficients are provided that are used to adjust the mapped values of spectral response acceleration for site response effects. These coefficients were developed based on observed site response characteristics in ground motion recordings from past earthquakes. Two coefficients are provided:

- The F_a coefficient is used to account for site response effects on short period ground shaking intensity.
- The F_v coefficient is used to account for site response effects of longer period motions.

Table 3.4 and Table 3.5 indicate the values of these coefficients as a function of site class and mapped MCE ground shaking acceleration values. Site-adjusted values of the MCE spectral response acceleration parameters at 0.2 and 1 s, respectively, are found from the following equations:

$$S_{MS} = F_a S_s \tag{3.1}$$

$$S_{M1} = F_v S_1 \tag{3.2}$$

The two site-adjusted spectral response acceleration parameters, S_{MS} and S_{M1}, permit a 5% damped, maximum considered earthquake ground shaking response spectrum to be constructed for the building site. This spectrum is constructed as indicated in Figure 3.11 and consists of a constant response acceleration range, between periods of T_0 and T_s, a constant response velocity range for periods in excess of T_s, and a short period range that ramps between an estimated zero period acceleration given by $S_{MS}/2.5$ and S_{MS}. Site-specific spectra can also be used. Regardless of whether site-specific spectra or

TABLE 3.5   Coefficient F_v as a Function of Site Class and Mapped Spectral Response Acceleration

| Site class | Mapped maximum considered earthquake spectral response acceleration at 1 s periods | | | | |
	$S_1 = 0.1$	$S_1 = 0.2$	$S_1 = 0.3$	$S_1 = 0.4$	$S_1 = 0.5$
A	0.8	0.8	0.8	0.8	0.8
B	1.0	1.0	1.0	1.0	1.0
C	1.7	1.6	1.5	1.4	1.3
D	2.4	2.0	1.8	1.6	1.5
E	3.5	3.2	2.8	2.4	a
F	a	a	a	a	a

Note: "a" indicates site-specific evaluation required.

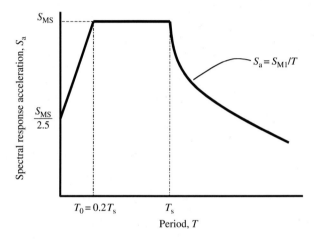

FIGURE 3.11   Maximum considered earthquake response spectrum.

spectra based on mapped values are used, the actual design values are taken as two thirds of the MCE values. The resulting design parameters are, respectively, labeled S_{DS} and S_{D1} and the design spectrum is identical to the MCE spectrum, except that the ordinates are taken as two thirds of the MCE values. The reason for using design values that are two thirds of the maximum considered values is that the design procedures, described in later sections, are believed to provide a minimum margin against collapse of 150%. Therefore, if design is conducted for two thirds of the MCE ground shaking, it is anticipated that buildings experiencing MCE ground shaking would be at incipient collapse, the desired performance objective for SUG-I structures.

3.5.4 Seismic Design Categories

The seismicity of the United States, and indeed the world, varies widely. It encompasses zones of very high seismicity in which highly destructive levels of ground shaking are anticipated to occur every 50 to 100 years and zones of much lower seismicity in which only moderate levels of ground shaking are ever anticipated. The *NEHRP Provisions* recognize that it is neither technically necessary nor economically appropriate to require the same levels of seismic protection for all buildings across these various regions of seismicity. Instead, the *NEHRP Provisions* assign each structure to a seismic design category (SDC) based on the level of seismicity at the building site, as represented by mapped shaking parameters, and the SUG.

Six SDCs, labeled A through F, are defined. SDC A represents the least severe seismic design condition and includes structures of ordinary occupancy located on sites anticipated to experience only very limited levels of ground shaking. SDC F represents the most severe design condition and includes

TABLE 3.6 Categorization of Structures into Seismic Design Category, Based on Design Short Period Spectral Response Acceleration, S_{DS} and Seismic Use Group

Value of S_{DS}	Seismic Use Group		
	I	II	III
$S_{DS} < 0.167g$	A	A	A
$0.167g \leq S_{DS} < 0.33g$	B	B	C
$0.33g \leq S_{DS} < 0.50g$	C	C	D
$0.50g \leq S_{DS}$	D^a	D^a	D^a

[a] *Seismic Use Group* I and II *structures* located on sites with mapped *maximum considered earthquake* spectral response acceleration at 1 s period, S_1, equal to or greater than $0.75g$ shall be assigned to *Seismic Design Category* E and *Seismic Use Group* III *structures* located on such sites shall be assigned to *Seismic Design Category* F.

TABLE 3.7 Categorization of Structures into Seismic Design Category, Based on Design One-Second Period Spectral Response Acceleration, S_{D1} and Seismic Use Group

Value of S_{D1}	Seismic Use Group		
	I	II	III
$S_{D1} < 0.067g$	A	A	A
$0.067g \leq S_{D1} < 0.133g$	B	B	C
$0.133g \leq S_{D1} < 0.20g$	C	C	D
$0.20g \leq S_{D1}$	D^a	D^a	D^a

[a] See footnote to Table 3.6.

structures assigned to SUG-III and located within a few kilometers of major, active faults, anticipated to produce very intense ground shaking. A designer determines to which SDC a structure should be assigned by reference to a pair of tables, reproduced as Table 3.6 and Table 3.7. A structure is assigned to the most severe category indicated by either table.

Nearly all aspects of the seismic design process are affected by the SDC that a structure is assigned to. This includes designation of the permissible structural systems, specification of required detailing, limitation on permissible heights and configuration, the types of analyses that may be used to determine the required lateral strength and stiffness, and the requirements for bracing and anchorage of non-structural components.

3.5.5 Permissible Structural Systems

The *NEHRP Provisions* define more than 70 individual seismic-force-resisting system types. These systems may be broadly categorized into five basic groups that include: bearing wall systems, building frame systems, moment-resisting frame systems, dual systems, and special systems:

- Bearing wall systems include those structures in which the vertical elements of the LFRS comprise either shear walls or braced frames in which the shear-resisting elements (walls or braces) are required to provide support for gravity (dead and live) loads in addition to providing lateral resistance. This is similar to the "box system" contained in earlier codes.
- Building frame systems include those structures in which the vertical elements of the LFRS comprise shear walls or braces, but in which the shear-resisting elements are not also required to provide support for gravity loads.
- Moment-resisting frame systems are those structures in which the lateral-force resistance is provided by the flexural rigidity and strength of beams and columns, which are interconnected in such a manner that stress is induced in the frame by lateral displacements.

- Dual systems rely on a combination of MRFs and either braced frames or shear walls. In dual systems, the braced frames or shear walls provide the primary lateral resistance and the MRF is provided as a back-up or redundant system, to provide supplemental lateral resistance in the event that earthquake response damages the primary lateral-force-resisting elements to an extent that they lose effectiveness.
- Special systems include unique structures, such as those that rely on the rigidity of cantilevered columns for their lateral resistance.

Within these broad categories, structural systems are further classified in accordance with the quality of detailing provided and the resulting ability of the structure to withstand earthquake-induced inelastic, cyclic demands. Structures that are provided with detailing believed capable of withstanding large cyclic inelastic demands are typically termed "special" systems. Structures that are provided with relatively little detailing and therefore, incapable of withstanding significant inelastic demands are termed "ordinary." Structures with limited levels of detailing and inelastic response capabilities are termed "intermediate." Thus, within a type of structure, for example, moment-resisting steel frames or reinforced concrete bearing walls, it is possible to have "special" MRFs or bearing walls, "intermediate" MRFs or shear walls, and "ordinary" MRFs or shear walls. The various combinations of such systems and construction materials results in a wide selection of structural systems to choose from. The use of "ordinary" and "intermediate" systems, regarded as having limited capacity to withstand cyclic inelastic demands, is generally limited to SDC A, B, and C and to certain low-rise structures in SDC D.

3.5.6 Design Coefficients

Under the *NEHRP Provisions*, required seismic design forces and, therefore, required lateral strength is typically determined by elastic methods of analysis, based on the elastic dynamic response of structures to design ground shaking. However, since most structures are anticipated to exhibit inelastic behavior when responding to the design ground motions, it is recognized that linear response analysis does not provide an accurate portrayal of the actual earthquake demands. Therefore, when linear analysis methods are employed, a series of design coefficients are used to adjust the computed elastic response values to suitable design values that consider probable inelastic response modification. Specifically, these coefficients are the response modification factor, R, the overstrength factor, Ω_0, and the deflection amplification coefficient, C_d. Tabulated values of these factors are assigned to a structure based on the selected structural system and the level of detailing employed in that structural system:

- The response modification coefficient, R, is used to reduce the required lateral strength of a structure, from that which would be required to resist the design ground motion in a linear manner to that required to limit inelastic behavior to acceptable levels, considering the characteristics of the selected structural system. Structural systems deemed capable of withstanding extensive inelastic behavior are assigned relatively high R values, as large as eight, permitting minimum design strengths that are only $\frac{1}{8}$ that required for elastic response to the design motion. Systems deemed to be incapable of providing reliable inelastic behavior are assigned low R values, approaching unity, requiring sufficient strength to resist design motion in a nearly elastic manner.
- The deflection amplification coefficient, C_d, is used to estimate the total elastic and inelastic lateral deformations of the structure when subjected to design earthquake ground motion. Specifically, lateral deflections calculated for elastic response of the structure to the design ground motion, reduced by the response modification coefficient R, are amplified by the factor C_d to obtain this estimate. The C_d coefficient accounts for the effects of viscous and hysteretic damping on structural response, as well as the effects of inelastic period lengthening. Structural systems that are deemed capable of developing significant amounts of viscous and hysteretic damping are assigned C_d values somewhat less than the value of the R coefficient. This results in an estimate of total lateral deformation that is somewhat lower than would be anticipated for a pure elastic response.

For structural systems with relatively poor capability to develop viscous and/or hysteretic damping, the C_d value may exceed R, resulting in estimates of lateral drift that exceed that calculated for elastic response.

- The overstrength coefficient, Ω_0, is used to provide an estimate of the maximum force likely to be delivered to an element in the structure, considering that due to effects of system and material overstrength this may be larger than the force calculated by elastic analysis of the structure's response to design ground motion, reduced by the response modification coefficient R. This overstrength factor is used to compute the required strength to resist behavioral modes that have limited capacity for inelastic response, such as column buckling or connection failure in braced frames.

Figure 3.12 illustrates the basic concepts behind these design coefficients. The figure contains an elastic design response spectrum, an elastic response line, and an inelastic response curve for an arbitrary structure, all plotted in lateral inertial force (base shear) versus lateral roof displacement coordinates. Response spectra are more familiarly plotted in coordinates of spectral response acceleration (S_a) versus structural period (T). It is possible to convert a spectrum plotted in that form to the spectrum shown in the figure through a two-step process. The first step consists of converting the response spectrum for S_a versus T coordinates to S_a versus spectral response displacement (S_d) coordinates. This is performed using the following relationship between S_a, S_d, and T:

$$S_d = \frac{T^2}{4\pi^2} S_a \tag{3.3}$$

Then the response spectrum is converted to the form shown in the figure by recognizing that for a structure responding in a given mode of excitation the base shear is equal to the product of the mass participation factor for that mode, the structure's mass, and the spectral response acceleration, S_a, at that period. Similarly, the lateral roof displacement for a structure responding in that mode is equal to the spectral response displacement times the modal participation factor. For a single degree of freedom structure, the mass participation factor and modal participation factor are both unity and the lateral base shear, V, is equal to the product of the spectral response acceleration at the mode of response and the mass of the structure, while the lateral roof displacement is equal to the spectral response displacement.

The dashed diagonal line in the figure represents the elastic response of the arbitrary structure. It is a straight line because a structure responding in an elastic manner will have constant stiffness and, therefore, a constant proportional relationship between the applied lateral force and resulting

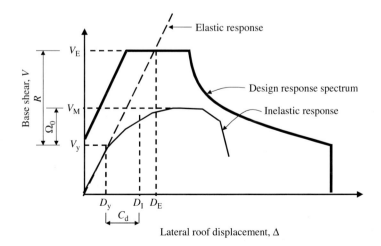

Figure 3.12 Schematic illustration of design coefficients.

displacement. The intersection of this diagonal line with the design response spectrum indicates the maximum total lateral base shear, V_E, and roof displacement, D_E, of the structure would develop if it responded to the design ground motion in an elastic manner. The third plot in the figure represents the inelastic response characteristics of this arbitrary structure, sometimes called a pushover curve. The pushover curve has an initial elastic region having the same stiffness as the elastic response line. The point V_y, D_y on the pushover curve represents the end of this region of elastic behavior. Beyond V_y, D_y, the curve is represented by a series of segments, with sequentially reduced stiffness, representing the effects of inelastic softening of the structure. The lateral base shear force, V_M, at the peak of the pushover curve, represents the maximum lateral force that the structure is capable of developing at full yield.

The response modification coefficient, R, is used in the provisions to set the minimum acceptable strength at which the structure will develop its first significant yielding, V_y. This is given by the simple relationship

$$V_y = \frac{V_E}{R} \tag{3.4}$$

The coefficient Ω_0 is used to approximate the full yield strength of the structure through the relationship

$$V_M = \Omega_0 V_y \tag{3.5}$$

The maximum total drift of the structure, D_I, is obtained from the relationship

$$D_I = C_d D_y \tag{3.6}$$

3.5.7 Analysis Procedures

The *NEHRP Provisions* permit the use of five different analytical procedures to determine the required lateral strength of a structure and to confirm that the structure has adequate stiffness to control lateral drift. The procedures permitted for a specific structure are dependent on the structure's SDC and its regularity.

3.5.7.1 Index Force Procedure

The index force procedure is permitted only for structures in SDC A. In this procedure, the structure must be designed to have sufficient strength to resist a static lateral force equal to 1% of the weight of the structure, applied simultaneously to each level. The forces must be applied independently in two orthogonal directions. Structures in SDC A are not anticipated ever to experience ground shaking of sufficient intensity to cause structural damage, provided that the structures are adequately tied together and have a complete LFRS. The nominal, 1%, lateral force function used in this procedure is intended as a means of ensuring that the structure has a complete LFRS of nominal, though somewhat arbitrary, strength. In addition to providing protection for the low levels of ground motion, anticipated for SDC A structures, this procedure is also considered to be a structural integrity provision, intended to provide nominal resistance against blast and other possible loading events.

3.5.7.2 Equivalent Lateral Force Analysis

The estimation of forces an earthquake may impose on a building can be accomplished by use of an equivalent lateral force (ELF) procedure. The ELF is commonly employed for simpler buildings and is provided in most building codes. The reader is referred to Hamburger (2002) for a review of building code and ELF development, but it will be noted that for many years the uniform building code (UBC) and other building codes determined the ELF according to variants on the equation $V = ZICKSW$, where V was the ELF or total design lateral force applied to the structure (or shear at the structure's base), determined from Z, a seismic zone factor varying between 0.075 (Zone 1, low-seismicity areas) and 0.40 (Zone 4, high-seismicity areas); I, an importance factor varying between 1.0 and 1.5; C, a function of the structure's fundamental period T, which effectively defined a response spectral shape; K, a factor varying by type of structure that accounted for the ductility of the structure and its material; S, a factor to account for site soil conditions; and W, the total seismic dead load.

In the 2000 *NEHRP Provisions*, ELF analysis may be used for any structure in SDC B and C, for any structure of light-frame construction, and for all regular structures, with a calculated structural period, T, not greater than $3.5T_s$, where T_s is as previously defined in Figure 3.11. ELF analysis consists of a simple approximation to modal response spectrum analysis. It only considers the first mode of a structure's lateral response and presumes that the mode shape for this first mode of response is represented by that of a simple shear beam. For structures having sufficiently low periods of first mode response ($T < 3.5T_s$) and regular vertical and horizontal distribution of stiffness and mass, this procedure approximates modal response spectrum analysis well. However, for longer period structures, higher mode response becomes significant and neglecting these higher modes results in significant errors in the estimation of structural response. Also, as the distribution of mass and stiffness in a structure becomes irregular, for example, the presence of torsional conditions or soft story conditions, the assumptions inherent in the procedure with regard to mode shape also become quite approximate, leading to errors. In SDCs D, E, and F, this method is permitted only for those structures where these inaccuracies are unlikely to be significant. The procedure is permitted for more general use in other SDCs because it is felt that the severity of design ground motion is low enough that inaccuracies in analysis of lateral response is unlikely to result in unacceptable structural performance and also because it is felt that designers in these regions of low seismicity may not be able to implement the more sophisticated and accurate methods properly.

As with the index force analysis procedure, the ELF consists of the simultaneous application of a series of static lateral forces to each level of the structure in each of the two independent orthogonal directions. In each direction, the total lateral force, known as the base shear, is given by the formula

$$V = \frac{S_{DS}}{R/I} W \tag{3.7}$$

This formula gives the maximum lateral inertial force that acts on an elastic, single degree of freedom structure with a period that falls within the constant response acceleration (periods shorter than T_s) portion of the design spectrum, reduced by the term R/I. In this formula, S_{DS} is the design spectral response acceleration at short periods, W is the dead weight of the structure and a portion of the supported live load, R is the response modification coefficient, and I is an occupancy importance factor, assigned based on the structure's SUG. For SUG-1 structures, I is assigned a value of unity. For SUG II and III structures, I is assigned values of 1.25 and 1.5, respectively. The effect of I is to reduce the permissible response modification factor, R, for structures in higher SUGs, requiring that the structures have greater strength, thereby limiting the permissible inelasticity and damage in these structures.

The base shear force given by Equation 3.7 need never exceed the following:

$$V = \frac{S_{D1}}{(R/I)T} W \tag{3.8}$$

Equation 3.8 represents the maximum lateral inertial force that acts on an elastic, single degree of freedom structure with period T that falls within the constant response velocity portion of the design spectrum (periods longer than T_s), reduced by the response modification coefficient, R and the occupancy importance factor, I. In this equation, all terms are as previously defined except that S_{D1} is the design spectral response acceleration at 1 s. For short period structures, Equation 3.7 will control. For structures with periods in excess of T_s, Equation 3.8 will control.

The shape of the design response spectrum shown in Figure 3.11 is not representative of the dynamic characteristics of ground motion found close to the fault rupture zone. Such motions are often dominated by a large velocity pulse and very large spectral displacement demands. Therefore, for structures in SDCs E and F, the seismic design categories for structures located close to major active faults, the base shear may not be taken less than the value given by Equation 3.9. Equation 3.9 approximates the effects of the additional long period displacements that have been recorded in some near field ground motion records

$$V = \frac{0.5S_1}{R/I} \tag{3.9}$$

The total, lateral base shear force given by Equations 3.7–3.9 must be distributed vertically for application to the various mass or diaphragm levels of the structure. For a structure with n levels, the force at diaphragm level x is given by the equation

$$F_x = C_{vx}V \tag{3.10}$$

where

$$C_{vx} = \frac{w_x h_x}{\sum_{i=1}^{n} w_i h_i} \tag{3.11}$$

h_x and h_i, respectively, are the heights of levels x and i above the structure's base. These formula are based on the assumption that the structure is responding in its first mode, in pure sinusoidal motion, and that the mode shape is linear. That is, it is assumed that at any instant of time, the displacement at level x of the structure is

$$\delta_x = \frac{h_x}{h_n}\delta_n \tag{3.12}$$

where δ_x and δ_n are the lateral displacements at level x and the roof of the structure, respectively, and h_n is the total height of the structure. For a structure responding in pure sinusoidal motion, the displacement δ_x, velocity v_x and acceleration a_x, of level x at any instant of time, t, is given by the following equations:

$$\delta_x = \delta_{x\max}\sin\left(\frac{2\pi}{T}t\right) \tag{3.13}$$

$$v_x = \delta_{x\max}\frac{2\pi}{T}\cos\left(\frac{2\pi}{T}t\right) \tag{3.14}$$

$$a_x = -\delta_{x\max}\frac{4\pi}{T^2}\sin\left(\frac{2\pi}{T}t\right) \tag{3.15}$$

Since acceleration at level x is directly proportional to the displacement at level x, the acceleration at level x in a structure responding in pure sinusoidal motion is given by the equation

$$a_x = \frac{h_n}{h_x}a_n \tag{3.16}$$

where a_n is the acceleration at the roof level. Since the inertial force at level x is equal to the product of mass at level x and the acceleration at level x, Equation 3.11 can be seen to be an accurate distribution of lateral inertial forces in a structure responding in a linear mode shape.

The lateral forces given by Equation 3.10 are applied to a structural model of the building and the resulting member forces and building interstory drifts are determined. The analysis must consider the relative rigidity of both the horizontal and vertical elements of the LFRS, and when torsional effects are significant, must consider three-dimensional distributions of stiffness, centers of mass, and rigidity. The structure must then satisfy two basic criteria. First, the elements of the LFRS must have sufficient strength to resist the calculated member forces in combination with other loads, and second, the structure must have sufficient strength to maintain computed interstory drifts within acceptable levels. The specific load combinations that must be used to evaluate member strength and the permissible interstory drifts are described in succeeding sections.

In recognition of the fact that higher mode participation can result in significantly larger forces at individual diaphragm levels, than is predicted by Equation 3.11, forces on diaphragms are computed using an alternative equation, as follows:

$$F_{px} = \frac{\sum_{i=x}^{n} F_i}{\sum_{i=x}^{n} w_i} w_{px} \tag{3.17}$$

where F_{px} is the design force applied to diaphragm level x, F_i is the force computed from Equation 3.11 at level i, w_{px} is the effective seismic weight, at level x, and w_i is the effective weight at level i.

3.5.7.3   Response Spectrum Analysis

Response spectrum analysis is permitted to be used for the design of any structure. The procedure contained in the *NEHRP Provisions* uses standard methods of elastic modal dynamic analysis, which are not described here, but are well documented in the literature, for example, by Chopra (1981). The analysis must include sufficient modes of vibration to capture participation of at least 90% of the structure's mass in each of the two orthogonal directions. The response spectrum used to characterize the loading on the structure may be either the generalized design spectrum for the site, shown in Figure 3.11, or a site-specific spectrum developed considering the regional seismic sources and site characteristics.

Regardless of the spectrum used, the ground motion is scaled by the factor (I/R), just as in the ELF technique. The *NEHRP Provisions* require that the member forces determined by response spectrum analysis be scaled so that the total applied lateral force in any direction be not less than 80% of the base shear calculated using the ELF method for regular structures nor 100% for irregular structures. This scaling requirement was introduced to ensure that assumptions used in building the analytical model does not result in excessively flexible representation of the structure and, consequently, an underestimate of the required strength.

3.5.7.4   Response History Analysis

Response history analysis is also permitted to be used for the design of any structure but, due to the added complexity, is seldom employed in practice except for special structures incorporating special base isolation or energy dissipation technologies. Either linear or nonlinear response history analysis is permitted to be used. When response history analysis is performed, input ground motion must consist of a suite of at least three pairs of orthogonal horizontal ground motion components, obtained from records of similar magnitude, source, distance, and site characteristics as the event controlling the hazard for the building's site. Each pair of orthogonal records must be scaled such that with a period range approximating the fundamental period of response of the structure, the square root of the sum of the squares of the orthogonal component ordinates envelopes 140% of the design response spectrum. Simple amplitude, rather than frequency domain scaling, is recommended. Actual records are preferred, though simulations may be used if a sufficient number of actual records representative of the design earthquake motion are not available. If a suite of less than seven records is used as input ground motion, the maximum of the response parameters (element forces and deformations) obtained from any of the records is used for design. If seven or more records are used, the mean values of the response parameters obtained from the suite of records may be used as design values. This requirement was introduced with the understanding that the individual characteristics of a ground motion record can produce significantly different results for some response quantities. It was hoped that this provision would encourage engineers to use larger suites of records and obtain an understanding of the variability associated with possible structural response.

When linear response history analyses are performed, the ground motion records, scaled as previously described, are further scaled by the quantity (I/R). The resulting member forces are combined with other loads, just as they would be if the ELF or response spectrum methods of analysis were performed.

When nonlinear response history analyses are performed, they must be used without further scaling. Rather than evaluating the strength of members using the standard load combinations considered with other analysis techniques, the engineer is required to demonstrate acceptable performance capability of the structure, given the predicted strength and deformation demands. The intention is that laboratory and other relevant data be used to demonstrate adequate behavior. This is a rudimentary introduction of performance-based design concepts, which will likely have significantly greater influence in future building codes.

3.5.8 Load Combinations and Strength Requirements

Structures must be proportioned with adequate strength to resist the forces predicted by the lateral seismic analysis together with forces produced by response to vertical components of ground shaking as well as dead and live loads. Unless nonlinear response history analysis is performed using ground motion records that include a vertical component of motion, the effects of vertical earthquake shaking are accounted for by the equation

$$E = Q_E \pm 0.2 S_{DS} D \tag{3.18}$$

where Q_E are the element forces predicted by the lateral seismic analysis, S_{DS} is the design spectral response acceleration at a 0.2 s response period, and D are the forces produced in the element by the structure's dead weight. The term $0.2 S_{DS} D$ represents the effect of vertical ground shaking response. For structures in zones of high seismicity, the term S_{DS} has a value approximating $1.0g$ and, therefore, the vertical earthquake effects are taken as approximately 20% increase or decrease in the dead load stress demands on each element. In fact, there are very few cases on record where structural collapse has been ascribed to the vertical response of a structure. This is probably because design criteria for vertical load resistance incorporate substantial factors of safety and also because most structures carry only a small fraction of their rated design live loads when they are subjected to earthquake effects. Therefore, most structures inherently have substantial reserve capacity to resist additional loading induced by vertical ground motion components. In recognition of this, most earlier codes neglected vertical earthquake effects. However, during the formulation of *ATC3.06*, it was felt to be important to acknowledge that ground shaking includes three orthogonal components. The resulting expression, which was somewhat arbitrary, ties vertical seismic forces to the short period design spectral response acceleration, as most structures are stiff vertically and have very short periods of structural response for vertical modes.

The earthquake forces on structural elements derived from Equation 3.18 are combined with dead and live loads in accordance with the standard strength level load combinations of *ASCE-7*. The pertinent load combinations are

$$Q = 1.4D \pm E \tag{3.19}$$

$$Q = 1.4D + 0.75(L + E) \tag{3.20}$$

where D, L, and E are the dead, live, and earthquake forces, respectively. Elements must then be designed to have adequate strength to resist these combined forces. The reduction factor of 0.75 on the combination of earthquake and live loads accounts for the low likelihood that a structure will be supporting full live load at the same time that it experiences full design earthquake shaking. An alternative set of load combinations is also available for use with design specifications that utilize allowable stress design formulations. These are essentially the same as Equations 3.19 and 3.20, except that the earthquake loads are further reduced by a factor of 1.4.

The *NEHRP Provisions* recognize that it is undesirable to allow some elements to experience inelastic behavior as they may be subject to brittle failure and in doing so compromise the ability of the structure to develop its intended inelastic response. The connections of braces to braced frames are an example of such elements. The *Provisions* also recognize that inelastic behavior in some elements, such as columns supporting discontinuous shear walls, could trigger progressive collapse of the structure. For these elements, the earthquake force *E* that must be used in the load combination (Equations 3.19 and 3.20) is given by the formula:

$$E = \Omega_0 Q_E \pm 0.2 S_{DS} D \tag{3.21}$$

where the term $0.2 S_{DS} D$ continues to represent the effects of vertical ground shaking response and the term $\Omega_0 Q_E$ represents an estimate of the maximum force likely to be developed in the element as a result of lateral earthquake response, considering the inelastic response characteristics of the entire structural system. In Equation 3.20, the term $\Omega_0 Q_E$ need never be taken larger than the predicted force on the element derived from a nonlinear analysis or plastic mechanism analysis.

3.5.9 Drift Limitations

It is important to control lateral drift in structures because excessive drift can result in extensive damage to cladding and other nonstructural building components. In addition, excessive lateral drift can result in the development of P–Δ instability and collapse.

Lateral drift is evaluated on a story by story basis. Story drift, δ, is computed as the difference in lateral deflection at the top of a story and that at the bottom of the story, as predicted by the lateral analysis. If the lateral analysis was other than a nonlinear response history analysis, design story drift, Δ, is obtained from the computed story drift, δ, by the equation:

$$\Delta = C_d \delta \tag{3.22}$$

where C_d is the design coefficient previously discussed. The design interstory drift computed from Equation 3.22 must be less than a permissible amount, dependent on the SUG and structural system as given in Table 3.8.

The provisions require evaluation of potential P–Δ instability through consideration of the quantity θ given by the equation:

$$\theta = \frac{P_x \Delta}{V_x h_x C_d} \tag{3.23}$$

In this equation, P_x is the dead weight of the structure above story x, Δ is the design story drift, computed from Equation 3.22, V_x is the design story shear obtained from the lateral force analysis, h_x is the story height, and C_d is the coefficient previously discussed. If the quantity θ computed by this equation is found to be less than 0.1, P–Δ effects may be neglected. If the quantity θ is greater than 0.1, P–Δ effects must be directly considered in performing the LFA. If the quantity θ exceeds 0.3, the structure should be considered potentially unstable and must be redesigned.

This approach to P–Δ evaluation has remained essentially unchanged since its initial introduction in *ATC3.06*. It was introduced in that document as a placeholder, pending the development of a more

TABLE 3.8 Permissible Drift Limits[a]

Structure	Seismic use group		
	I	II	III
Structures other than masonry *shear wall* or masonry *wall*-frame *structures*, four stories or less in height with interior *walls*, partitions, ceilings, and exterior *wall* systems that have been designed to accommodate the *story* drifts	$0.025 h_{sx}$[b]	$0.020 h_{sx}$	$0.015 h_{sx}$
Masonry cantilever *shear wall structures*[c]	$0.010 h_{sx}$	$0.010 h_{sx}$	$0.010 h_{sx}$
Other masonry *shear wall structures*	$0.007 h_{sx}$	$0.007 h_{sx}$	$0.007 h_{sx}$
Masonry *wall*-frame *structures*	$0.013 h_{sx}$	$0.013 h_{sx}$	$0.010 h_{sx}$
All other *structures*	$0.020 h_{sx}$	$0.015 h_{sx}$	$0.010 h_{sx}$

[a] There shall be no drift limit for single-story structures with interior walls, partitions, ceilings, and exterior wall systems that have been designed to accommodate the story drifts.

[b] h_{sx} is the story height below Level *x*.

[c] Structures in which the basic structural system consists of masonry shear walls designed as vertical elements cantilevered from their base or foundation support which are so constructed that moment transfer between shear walls (coupling) is negligible.

accurate method for evaluating drift-induced instability. Obvious deficiencies in this current approach include the fact that it evaluates drift effects at the somewhat artificial design-base shear levels. A more realistic evaluation would consider the actual expected lateral deformations of the structure as well as the yield level shear capacity of the structure at each story. As contained in the current provisions, evaluation of P–Δ effects seldom controls a structure's design.

3.5.10 Structural Detailing

Structural detailing is a critical feature of seismic-resistant design but is not generally specified by the *NEHRP Provisions*. Rather, the Provisions adopt detailing requirements contained in standard design specifications developed by the various materials industry associations including the American Institute of Steel Construction, the American Concrete Institute, the American Forest Products Association, and the Masonry Society. Other chapters in this handbook present the requirements of these various design standards.

Glossary

Attenuation — The rate at which earthquake ground motion decreases with distance.

Base shear — The total lateral force for which a structure is designed using equivalent lateral force techniques.

Characteristic earthquake — A relatively narrow range of magnitudes at or near the maximum that can be produced by the geometry, mechanical properties, and state of stress of a fault (Schwartz and Coppersmith 1987).

Completeness — Homogeneity of the seismicity record.

Cripple wall — A carpenter's term indicating a wood frame wall of less than full height T, usually built without bracing.

Critical damping — The value of damping such that free vibration of a structure will cease after one cycle (ccrit = 2 $m\omega$).

Damage — Permanent, cracking, yielding, or buckling of a structural element or structural assemblage.

Damping — Energy dissipation that occurs in a dynamically deforming structure, either as a result of frictional forces or structural yielding. Increased damping tends to reduce the amount that a structure responds to ground shaking.

Degradation — A behavioral mode in which structural stiffness or strength is reduced as a result of inelastic behavior.

Design (basis) earthquake — The earthquake (as defined by various parameters, such as PGA, response spectra, etc.) for which the structure will be, or was, designed.

Ductile detailing — Special requirements, such as for reinforced concrete and masonry, close spacing of lateral reinforcement to attain confinement of a concrete core, appropriate relative dimensioning of beams and columns, 135° hooks on lateral reinforcement, hooks on main beam reinforcement within the column, etc.

Ductile frames — Frames required to furnish satisfactory load-carrying performance under large deflections (i.e., ductility). In reinforced concrete and masonry this is achieved by ductile detailing.

Ductility factor — The ratio of the total displacement (elastic plus inelastic) to the elastic (i.e., yield) displacement.

Elastic — A mode of structural behavior in which a structure displaced by a force will return to its original state upon release of the force.

Fault — A zone of the earth's crust within which the two sides have moved — faults may be hundreds of miles long — from one to over one hundred miles deep, and not readily apparent on the ground surface.

Ground shaking — A random, rapid cyclic motion of the ground produced by an earthquake.

Hysteresis — A form of energy dissipation that is related to inelastic deformation of a structure.

Inelastic — A mode of structural behavior in which a structure, displaced by a force, exhibits permanent unrecoverable deformation.

Lateral force resisting system (LFRS) — A structural system for resisting horizontal forces due, for example, to earthquake or wind (as opposed to the vertical force resisting system, which provides support against gravity).

Liquefaction — A process resulting in a soil's loss of shear strength due to a transient excess of pore water pressure.

Magnitude — A unique measure of an individual earthquake's release of strain energy, measured on a variety of scales, of which the moment magnitude M_w (derived from seismic moment) is preferred.

Mass participation — That portion of total mass of a multidegree of freedom structure that is effective in a given mode of response.

MCE — Maximum considered earthquake — the earthquake intensity forming the basis for design in the *NEHRP Provisions.*

Mode shape — A deformed shape in which a structure can oscillate freely when displaced.

Natural mode — A characteristic dynamic property of a structure in which it will oscillate freely.

Nonductile frames — Frames lacking ducility or energy absorption capacity due to lack of ductile detailing — ultimate load is sustained over a smaller deflection (relative to ductile frames) and for fewer cycles.

Participation factor — A mathematical relationship between the maximum displacement of a multi-degree of freedom structure and a single degree of freedom structure.

Peak ground acceleration (PGA) — The maximum amplitude of recorded acceleration (also termed the ZPA, or zero period acceleration).

Period — The amount of time it takes a structure that has been displaced in a particular natural mode and then released to undergo one complete cycle of motion.

Pounding — The collision of adjacent buildings during an earthquake due to insufficient lateral clearance.

Response spectrum — A plot of maximum amplitudes (acceleration, velocity, or displacement) of a single degree of freedom oscillator (sdof), as the natural period of the sdof is varied across a spectrum of engineering interest (typically, for natural periods from 0.03 to 3 or more seconds, or frequencies of 0.3 to 30+ Hz).

Reverse fault — A fault that exhibits dip-slip motion, where the two sides are in compression and move away toward each other.

Seismic risk — The product of the hazard and the vulnerability (i.e., the expected damage or loss, or the full probability distribution).

Soft story — A story of a building signifiantly less stiff than adjacent stories (i.e., the lateral stiffness is 70% or less than that in the story above, or less than 80% of the average stiffness of the three stories above; BSSC 1994).

Spectral acceleration — The maximum response acceleration that a structure of given period will experience when subjected to a specific ground motion.

Spectral displacement — The maximum response displacement that a structure of given period will experience when subjected to a specific ground motion.

Spectral velocity — The maximum response velocity that a structure of given period will experience when subjected to a specific ground motion.

Spectrum amplification factor — The ratio of a response spectral parameter to the ground motion parameter (where parameter indicates acceleration, velocity, or displacement).

Viscous — A form of energy dissipation that is proportional to velocity.

Yielding — A behavioral mode in which a structural displacement increases under application of constant load.

References

Aiken, I.D., Nims, D.K., Whittaker, A.S., and Kelly, J.M. 1993. Testing of Passive Energy Dissipation Systems. *Earthquake Spectra*, Volume 9, number 3, August, pp. 336–370.

Building Seismic Safety Council. 1997. *NEHRP Recommended Provisions for Seismic Regulations for Buildings and Other Structures*. Report No. FEMA 302/303. Federal Emergency Management Agency, Washington, DC.

Building Seismic Safety Council. 2001. *NEHRP Recommended Provisions for Seismic Regulations for Buildings and Other Structures, 2000 Edition*. Report no. FEMA 368. Federal Emergency Management Agency, Washington, DC.

Chen, W.F. and Scawthorn, C. (eds). 2002. *Earthquake Engineering Handbook*, CRC Press, Boca Raton, FL.

Dowrick, D.J. 1987. *Earthquake Resistant Design*, 2nd edition, John Wiley & Sons, New York.

Earthquake Resistant Design Codes in Japan, January 2000. Japan Society of Civil Engineers, Tokyo.

Federal Emergency Management Agency. 1988. *Rapid Visual Screening of Buildings for Potential Seismic Hazards: A Handbook*, FEMA 154, FEMA, Washington, DC.

Hamburger, R.O. 2002. *Building Code Provisions for Seismic Resistance*, chapter in Chen and Scawthorn (2002).

Hanson, R.D., Aiken, I.D., Nims, D.K., Richter, P.J., and Bachman, R.E. 1993. *State-of-the-Art and State-of-the-Practice in Seismic Energy Dissipation*. Proceedings of ATC-17-1 Seminar on Seismic Isolation, Passive Energy Dissipation, and Active Control; San Francisco, California, March 11–12, 1993, Volume 2: Passive Energy Dissipation, Active Control, and Hybrid Control Systems. Applied Technology Council, Redwood City, CA, pp. 449–471.

IAEE. 1996. *Regulations for Seismic Design: A World List, 1996 (RSD)*. Rev. ed. (update of Earthquake Resistant Regulations: A World List, 1992). Prepared by the International Association for Earthquake Engineering (IAEE), Tokyo, 1996; and its Supplement 2000: Additions to Regulations for Seismic Design: A World List, 1996. Available from Gakujutsu Bunken Fukyu-Kai (Association for Science Documents Information), c/o Tokyo Institute of Technology, 2-12-1 Oh-Okayama, Meguro-Ku, Tokyo, Japan 152-8550 (telephone: +81-3-3726-3117; fax: +81-3-3726-3118; e-mail: gakujyutu-bunken@mvd.biglobe.ne.jp).

ICC. 2000. *International Building Code 2000*, International Code Council, published by International Conference of Building Officials, Whittier, CA, and others.

Iemura, H. and Pradono, M.H. 2002. *Structural Control*, chapter in Chen and Scawthorn (2002).

Lou, J.Y.K., Lutes, L.D., and Li, J.J. 1994. *Active Tuned Liquid Damper for Structural Control*. Proceedings [of the] First World Conference on Structural Control: International Association for Structural Control; Los Angeles, Augusts 3–5, 1994; Housner, G.-W. et al., eds; International Association for Structural Control, Los Angeles, Volume 2, pp. TP1-70–TP1-79.

Naeim, F. (ed.). 2001. *Seismic Design Handbook*, 2nd edition, Kluwer, New York.

NFPA. n.d. *NFPA 5000 Building Code*, National Fire Protection Association, Cambridge, MA (publication pending).

Paz, M. (ed.). 1994. *International Handbook of Earthquake Engineering: Codes, Programs, and Examples*, Chapman & Hall, New York.

Perry, C.L., Fierro, E.A., Sedarat, H., and Scholl, R.E. 1993. Seismic Upgrade in San Francisco Using Energy Dissipation Devices. *Earthquake Spectra*, Volume 9, number 3, August, pp. 559–580.

Soong, T.T. and Costantinou, M.C. (eds.). 1994. *Passive and Active Structural Vibration Control in Civil Engineering*, State University of New York, Buffalo, NY (CISM International Centre for Mechanical Sciences, Vol. 345), VIII, 380 pp. Springer-Verlag, Vienna, New York.

Spencer Jr., B.F., Sain, M.K., Won, C.-H. Kaspari Jr., D.C., and Sain, P.M. 1994. Reliability-Based Measures of Structural Control Robustness, *Struct. Safety*, Volume 15, pp. 111–129.

Structural Engineers Association of California. 1996. *Vision 2000, A Framework for Performance-based Seismic Design*, Structural Engineers Association of California, Sacramento, CA.

Structural Engineers Association of California. 1999. Seismology Committee. *Recommended Lateral Force Requirements and Commentary*. Sacramento, CA.

U.S. Army. 1992. *Seismic Design for Buildings* (Army: TM 5-809-10; Navy: NAVFAC P-355; USAF: AFM-88-3, Chapter 13). Washington, DC: Departments of the Army, Navy, and Air Force, 1992. 407 pages. Available from the National Technical Information Service (NTIS), Military Publications Division, Washington.

Uniform Building Code. 1997. *Volume 2, Structural Engineering Design Provisions*, International Conference on Building Officials, Whittier.

Yang, Y.-B., Chang, K.-C., and Yau, J.-D. 2002. *Base Isolation*, chapter in Chen and Scawthorn (2002).

Further Reading

Chen and Scawthorn (2002) provide an extensive reference on earthquake engineering, while Naiem (2001) provides an excellent resource on seismic design. Both references have individual chapters on design of steel, wood, reinforced concrete, reinforced masonry, and precast structures, and also on nonstructural elements. SEAOC (1999), BSSC (1997), and SEAOC (1996) provide an excellent overview of the current state of seismic design requirements. U.S. Army (1992) and Dowrick (1987) are also useful, although a bit older. Some useful sources on seismic code provisions in countries other than the U.S. include *Earthquake Resistant Design Codes in Japan* (2000), Paz (1995), and IAEE (1996). The last is a comprehensive compendium of seismic regulations for over 40 countries, including *Eurocode 8* (the European Union's seismic provisions).

4

Seismic Design of Bridges*

Lian Duan
*Division of Engineering Services,
California Department of
Transportation,
Sacramento, CA*

Mark Reno
*Quincy Engineering,
Sacramento, CA*

Wai-Fah Chen
*College of Engineering,
University of Hawaii at Manoa,
Honolulu, HI*

Shigeki Unjoh
*Ministry of Construction,
Public Works Research Institute,
Tsukuba, Ibaraki, Japan*

4.1   Introduction

Bridges are very important elements in the modern transportation system. Recent earthquakes, particularly the 1989 Loma Prieta and the 1994 Northridge Earthquakes in California, the 1995 Hyogo-Ken Nanbu Earthquake in Japan, the 1999 JiJi Earthquake in Taiwan, and the 1999 Kocaeli Earthquake in Turkey, have caused collapse of, or severe damage to, a considerable number of major bridges [1,2]. Since the 1989 Loma Prieta Earthquake in California [3], extensive research [4–18] has been conducted on seismic design and retrofit of bridges in Japan and the United States, especially in California.

---

*Much of the material of this chapter was taken from Duan, L. and Chen, W.F., Chapter 19: Bridges, in Earthquake Engineering Handbook, Chen, W.F. and Scawthorn, C., Ed., CRC Press, Boca Raton, FL, 2002.

This chapter first addresses the seismic bridge design philosophies and conceptual design in general, then discusses mainly the U.S. seismic design practice to illustrate the process, and finally presents briefly seismic design practice in Japan.

4.2 Earthquake Damages to Bridges

Past earthquakes have shown that the damage induced in bridges can take many forms depending on the ground motion, site conditions, structural configuration, and specific details of the bridge [1]. Damage within the superstructure has rarely been the primary cause of collapse. Most of the severe damage to bridges has taken one of the following forms [1]:

- *Unseating of the superstructure at in-span hinges or simple supports due to inadequate seat lengths or restraint.* A skewed, curved, or complex superstructure framing configuration further increases the vulnerability. Figure 4.1 shows the collapsed upper and lower decks of the eastern portion of the San Francisco-Oakland Bay Bridge (SFOBB) in the 1989 Loma Prieta Earthquake, which can be attributed to anchor bolt failures allowing the span to move. Figure 4.2 shows collapsed I-5 Gavin Canyon Undercrossing, California, in the 1994 Northridge Earthquake, which can be attributed to geometric complexities arising from 66° skew angle abutments, in-span expansion joints, as well as the inadequate 300 mm seat width. For simply supported bridges, these failures are most likely when ground failure induces relative motion between the spans and their supports.
- *Column brittle failure due to deficiencies in shear capacity and inadequate ductility.* In reinforced concrete columns, the shear capacity and ductility concerns usually stem from inadequate lateral and confinement reinforcement. Figure 4.3 shows the collapsed 600 m of Hanshin Expressway in the 1995 Hyogo-Ken Nanbu Earthquake in Japan where failure was attributed to deficiencies in shear design and poor ductility. In steel columns, the inadequate ductility usually stems from progressive local buckling, fracture, and global buckling leading to collapse.
- *Unique failures in complex structures.* Figure 4.4 shows the collapsed Cypress Street Viaduct, California, in the 1989 Loma Prieta Earthquake where the unique vulnerability was the inadequately reinforced pedestal above the first level. In outrigger column bents, the vulnerability may be in the cross-beam or the beam–column joint.

FIGURE 4.1 Collapse of eastern portion of San Francisco-Oakland Bay Bridge in the Loma Prieta Earthquake (courtesy of California Department of Transportation).

FIGURE 4.2 Collapse of I-5 Gavin Canyon Undercrossing, California, in the 1994 Northridge Earthquake.

FIGURE 4.3 Collapse of Hanshin Expressway in the 1995 Hyogo-Ken Nanbu Earthquake in Japan (courtesy of Mark Yashinsky).

4.3 Seismic Design Philosophies

4.3.1 Design Evolution

Seismic bridge design has been improving and advancing, based on research findings and lessons learned from past earthquakes. In the United States, prior to the 1971 San Fernando Earthquake, the seismic design of highway bridges was partially based on lateral force requirements for buildings.

FIGURE 4.4   Collapse of Cypress Viaduct, California, in the 1989 Loma Prieta Earthquake.

Lateral loads were considered as levels of 2 to 6% of dead loads. In 1973, California Department of Transportation (Caltrans) developed new seismic design criteria (SDC) related to site, seismic response of the soils at the site, and dynamic characteristics of bridges. The American Association of State Highway and Transportation Officials (AASHTO) modified the Caltrans 1973 Provisions slightly and adopted Interim Specifications. Applied Technology Council (ATC) developed guidelines ATC-6 [19] for seismic design of bridges in 1981. AASHTO adopted ATC-6 as the Guide Specifications in 1983 and in 1991 incorporated it into the Standard Specifications for Highway Bridges [20].

Prior to the 1989 Loma Prieta Earthquake, bridges in California were typically designed using a single-level force-based design approach based on a "no-collapse" design philosophy. Seismic loads were determined based on a set of soil conditions and a suite of four site-based standard acceleration response spectra (ARS). Structures were analyzed using the three-dimensional elastic dynamic multimodal response spectrum analysis method. Structural components were designed to resist forces from the response spectrum analysis that were modified with a "Z-factor." The "Z-factor" was based on the individual structural element ductility and degree of risk. Minimum transverse reinforcement levels were required to meet confinement criteria [21].

Since 1989, the design criteria specified in Caltrans design manuals [21–27] have been updated continuously to reflect recent research findings and development in the field of seismic bridge design. Caltrans has been shifting toward a displacement-based design approach emphasizing element and system capacity design. For important bridges in California the performance-based project-specific design criteria [28–30] have been developed since 1989.

FHWA updated its *Seismic Design and Retrofit Manual for Highway Bridges* in 1995 [31,32]. ATC published the improved SDC recommendations for California bridges [33] in 1996 and for U.S. bridges and highway structures [34] in 1997. Caltrans published the performance and displacement-based Seismic Design Criteria Version 1.3 [35] in 2004, which focuses mainly on concrete bridges, and the Guide Specifications for Seismic Design of Steel Bridge (*Guide*) [36] in 2001. Most recently, the NCHRP 12-49 team developed a new set of LRFD Guidelines (*Guidelines*) for the seismic design of highway bridges [37], compatible with the AASHTO-LFRD Bridge Design Specifications [38]. Significant advances in earthquake engineering were made during the last decade of the twentieth century.

4.3.2 No-Collapse-Based Design

The basic design philosophy is to prevent bridges from collapsing during severe earthquakes [24–27,35–37] that have only a small probability of occurring during the useful life of the bridge. To prevent collapse, two alternative approaches are commonly used in design. The first is a conventional force-based approach where the adjustment factor Z for ductility and risk assessment [21], or the response modification factor R [20,37], is applied to elastic member forces obtained from a response spectra analysis or an equivalent static analysis. The second approach is a more recent displacement-based approach [24,35] where displacements are a major consideration in design. For more detailed information, references can be made to comprehensive discussions in *Seismic Design and Retrofit of Bridges* by Prietley et al. [16], *Bridge Engineering Handbook* by Chen and Duan [39], and Refs [40,41].

4.3.3 Performance-Based Design

Following the 1989 Loma Prieta Earthquake, bridge engineers [3] have faced three essential challenges:

- Ensure that earthquake risks posed by new construction are acceptable.
- Identify and correct unacceptable seismic safety conditions in existing structures.
- Develop and implement the rapid, effective, and economic response mechanism for the recovering structural integrity after damaging earthquakes.

Performance-based project-specific criteria [28–30] and design memoranda [24] have been developed and implemented for the design and retrofitting of important bridges by California bridge engineers. These performance-based criteria included guidelines for development of site-specific ground motion estimates, ductile design details to preclude brittle failure modes, rational procedures for concrete joint shear design, and the definition of limit states for various performance objectives [15]. The performance-based criteria usually require a two-level design approach. The first level of design is to ensure the performance (service) of a bridge in small-magnitude earthquake events that may occur several times during the life of the bridge. The second level of design is to achieve the performance (no collapse) of a bridge under severe earthquakes that have only a small probability of occurring during the useful life of the bridge. Figure 4.5 shows a flowchart for development of performance-based SDC.

4.4 Seismic Conceptual Design

Bridge design is a complex engineering process involving consideration of numerous important factors, such as bridge structural systems, materials, dimensions, foundations, esthetics, and local landscape and surrounding environment [16,42]. Selecting an appropriate earthquake-resisting system (ERS) to resolve the potential conflicts between the configuration and seismic performance should be completed as early as possible in the design effort. For a desirable seismic resistant design, the following guidelines may be useful:

- Bridge type, component and member dimensions, and esthetics shall be investigated to reduce the seismic demands to the greatest extent possible. Esthetics should not be the primary reason for producing undesirable frame and component geometry.
- Bridges should ideally be as straight as possible. Curved bridges complicate and potentially magnify seismic responses.
- Superstructures should be continuous with as few joints as possible. Necessary restrainers and sufficient seat width shall be provided between adjacent frames at all expansion joints, and at the seat-type abutments to eliminate the possibility of unseating during a seismic event.
- Support skew angles should be as small as possible, that is, abutments and piers should be oriented as close to perpendicular to the bridge longitudinal axis as practical (within 20 to 25°) even at the expense of slightly increasing the bridge length. Highly skewed abutments and piers are vulnerable to damage from undesired rotational response and increased seismic displacement demands.

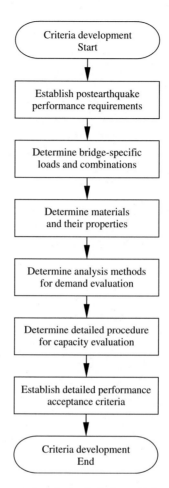

FIGURE 4.5   Development of performance-based seismic design criteria.

- Adjacent frames or piers should be proportioned to minimize the differences in the fundamental periods and skew angles, and to avoid drastic changes in stiffness and strength in both the longitudinal and transverse directions. Dramatic changes in stiffness can result in damage to the stiffer frames or piers. It is strongly recommended [35] that the effective stiffness between any two bents within a frame, or between any two columns within a bent, not vary by a factor of more than two. Similarly, it is highly recommended that the ratio of the shorter fundamental period to the longer fundamental period for adjacent frame in the longitudinal and transverse directions be larger than 0.7.
- Structural configurations that cannot accommodate the recommendations must be capable of accommodating the associated large relative displacement without compromising structural integrity. Each frame shall provide a well-defined load path with predetermined plastic hinge locations and utilize redundancy whenever possible. Balanced mass and stiffness distribution (Figure 4.6) in a frame results in a structure response that is more predictable and is more likely to respond in its fundamental mode of vibration. Simple analysis tools can then be used to predict the structures response with relative accuracy, whereas irregularities in geometry increase the likelihood of complex nonlinear response that is difficult to accurately predict by elastic modeling or plane frame inelastic static analysis. The following various techniques may be used to achieve balanced geometry to create a uniform and more predictable structure response [43]:
  a. Adjust foundation rotational and translation stiffness (e.g., use oversized pile shafts).
  b. Adjust effective column lengths (e.g., lower footings, provide isolation casings).

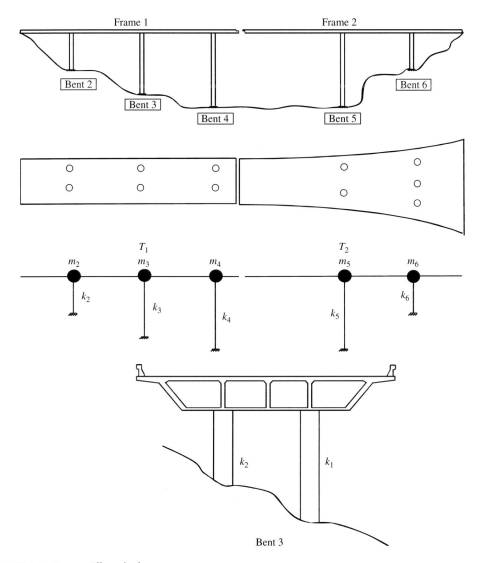

FIGURE 4.6 Frame stiffness [35].

c. Modify end fixities.

d. Reduce/redistribute superstructure mass.

e. Vary the column cross-section and longitudinal reinforcement ratios.

f. Add or relocate columns/piers.

g. Modify the hinge/expansion joint layout.

In the event of other constraints preventing the designer from achieving balance between frames of a bridge, the following recommendations may be considered [35]:

a. Isolate adjacent frames longitudinally by providing a large expansion gap to reduce the likelihood of pounding.

b. Provide adequate seat width to prevent unseating at hinges. Seat extenders may be used; however, they should be isolated transversely to avoid transmitting large lateral shear forces between frames.

c. Limit the transverse shear capacity between frames to prevent large lateral forces from being transferred to the stiffer frame.

      d. Avoid placing hinge seats between unbalanced frames by placing expansion joints between frames with short cantilever spans that butt up to one another.

- Seismic protective devices, that is, energy dissipation and isolation devices may be provided at appropriate locations, thereby reducing the seismic force effects. The energy dissipation devices are to increase the effective damping of the structure thereby reducing forces, deflections, and impact loads. Isolation devices are to lengthen the fundamental mode of vibration and to isolate potentially large superstructure inertial loads from the substructure so that structure is subject to lower earthquake forces.

- For concrete bridges, structural components shall be proportioned to direct inelastic damage into the columns, pier walls, and abutments. The superstructure shall have sufficient overstrength to remain essentially elastic if the columns/piers reach their most probable plastic moment capacity. The superstructure-to-substructure connection for nonintegral caps may be designed to fuse prior to generating inelastic response in the superstructure. Concrete columns shall be well proportioned, moderately reinforced, and easily constructed. The girders, bent caps, and columns shall be proportioned to minimize joint stresses. Moment-resisting connections shall have sufficient joint shear capacity to transfer the maximum plastic moments and shears without joint distress.

- Initial sizing of columns should be based on slenderness ratios, bent cap depth, compressive dead-to-live load ratio, and service loads. Columns shall demonstrate dependable postyield displacement capacity without an appreciable loss of strength. Thrust–moment–curvature (P–M–Φ) relationships should be used to verify a column's satisfactory performance under service and seismic loads. Abrupt changes in the cross-section and the bending capacity of columns should be avoided. Columns must have sufficient rotation capacity to achieve the target displacement ductility requirements.

- For steel bridges, structural components shall be generally designed to ensure that inelastic deformation only occur in the specially detailed ductile substructure elements. Inelastic behavior in the form of controlled damage may be permitted in some of the superstructure components such as end cross-frames or diaphragms, shear keys, and bearings. The inertial forces generated by the deck must be transferred to the substructure through girders, trusses, cross-frames, lateral bracings, end diaphragms, shear keys, and bearings. As an alternative, specially designed ductile end-diaphragms may be used as structural fuses to reduce potential damage in other parts of the structure.

- Steel multicolumn bents or towers shall be designed as ductile moment-resisting frames (MRFs) or ductile braced frames such as concentrically braced frames (CBFs) and eccentrically braced frames (EBFs). For components expected to behave inelastically, elastic buckling (local compression and shear, global flexural, and lateral torsion) and fracture failure modes shall be avoided. All connections and joints should preferably be designed to remain essentially elastic. For MRFs, the primary inelastic deformation shall preferably be columns. For CBFs, diagonal members shall be designed to yield when members are in tension and to buckle inelastically when they are in compression. For EBFs, a short beam segment designated as a "link" shall be well designed and detailed to provide ductile structural behavior.

- The ATC/MCEER recommended LRFD *Guidelines* [37] classify the ERS into permissible and not recommended categories (Figures 4.7 to 4.11) based on consideration of the most desirable seismic performance ensuring wherever possible postearthquake serviceability. Figure 4.12 shows design approaches for the permissible ERS.

4.5   Seismic Performance Criteria

4.5.1   ATC/MCEER Guidelines

Table 4.1 gives the seismic performance criteria for highway bridges specified in the proposed ATC/MCEER *Guidelines* [37]. As a minimum, a bridge shall be designed for the life safety level of

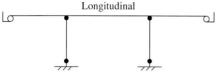

Abutment resistance not
required as part of ERS

Plastic hinges in inspectable locations
or elastic design of columns

Knock-off backwalls permissible

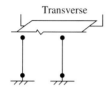

Abutment not required in ERS, breakaway
shear keys permissible

Plastic hinges in inspectable locations
or elastic design of columns

Abutment resistance required, but abutment
able to resist 3% in 75-year earthquake elastically
and passive soil pressure in longitudinal direction
is less than 0.70 × presumptive value given in
Article 7.5.2 in *Guidelines*

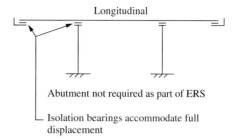

Abutment not required as part of ERS

Isolation bearings accommodate full
displacement

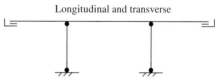

Isolation bearings with significant energy
dissipation capacity or energy dissipators are
used at the abutment to limit overall
displacements

Plastic hinges in inspectable locations
or elastic design of columns

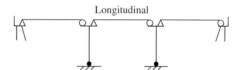

Multiple simply supported spans with adequate
seat widths. Plastic hinges in inspectable locations
or elastic design of columns

FIGURE 4.7   Permissible earthquake resisting systems [37].

performance. Higher level of performance may be required depending upon the bridge's importance and owner's requirements.

The seismic performance criteria shown in Table 4.1 shall be achieved by the following design objectives:

- *Columns as primary energy dissipation mechanism.* The main objective is to force the inelastic deformations to occur primarily in the columns in order that the earthquake damage can be easily inspected and readily repaired after an earthquake. The amount of longitudinal steel in the reinforced columns should be minimized to reduce foundation and connection costs.
- *Abutments as an additional energy dissipation mechanism.* The objective is to expect the inelastic deformations to occur in the columns as well as the abutments in order to either minimize column size and reduce ductility demand on the column.
- *Isolation bearings as main energy dissipation mechanism.* The objective is to lengthen the period of a relatively stiff bridge and which results in a lower design seismic force. Energy dissipation will occur in the isolation bearings and columns are usually then expected to perform elastically.
- *Structural components between deck and columns/abutments as energy dissipation mechanism.* The objective is to design ductile components that do not result in reduced design force but will reduce the ductility demands on the columns in order to minimize the energy that is dissipated in the plastic hinge zone of columns.

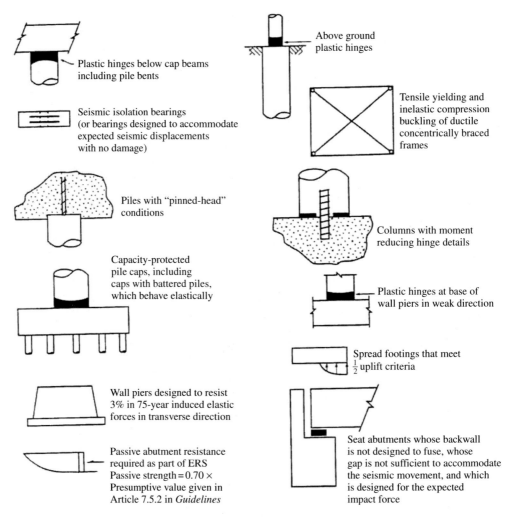

FIGURE 4.8 Permissible earthquake resisting elements [37].

- *Replaceable/renewable sacrificial plastic hinge elements as energy dissipation mechanism.* The objective is to control damage and permit significant inelastic deformation to occur at a specially designed replaceable/renewable sacrificial plastic hinge elements in the plastic hinge zone of a column. The concept is similar to the conventional ductile design concept that permits significant inelastic deformation in the plastic hinge zone of a column. The difference compared with the conventional ductile design is that construction details in the plastic hinge zone of concrete columns provide a replaceable/renewable sacrificial plastic hinge elements. The concept has been extensively tested [44] but has not been used in practice.

4.5.2 Caltrans

Table 4.2 outlines Caltrans seismic performance criteria [24] including the bridge classification, the service, and damage levels established in 1994 [15]. A bridge is categorized as "Important" or "Ordinary."

For Standard "Ordinary" bridges, the displacement-based one-level safety-evaluation design ("no-collapse" design) is only required in the Caltrans SDC [35]. Nonstandard "Ordinary" bridges feature irregular geometry and framing (multilevel, variable width, bifurcating, or highly horizontally

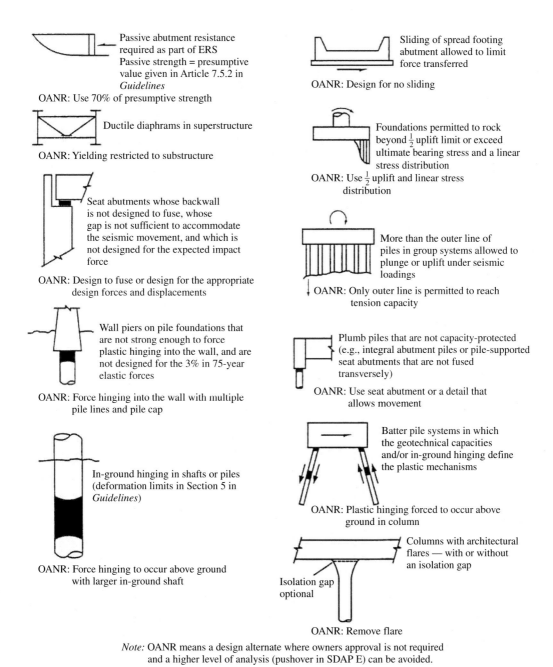

Passive abutment resistance required as part of ERS Passive strength = presumptive value given in Article 7.5.2 in *Guidelines*
OANR: Use 70% of presumptive strength

Ductile diaphrams in superstructure
OANR: Yielding restricted to substructure

Seat abutments whose backwall is not designed to fuse, whose gap is not sufficient to accommodate the seismic movement, and which is not designed for the expected impact force
OANR: Design to fuse or design for the appropriate design forces and displacements

Wall piers on pile foundations that are not strong enough to force plastic hinging into the wall, and are not designed for the 3% in 75-year elastic forces
OANR: Force hinging into the wall with multiple pile lines and pile cap

In-ground hinging in shafts or piles (deformation limits in Section 5 in *Guidelines*)
OANR: Force hinging to occur above ground with larger in-ground shaft

Sliding of spread footing abutment allowed to limit force transferred
OANR: Design for no sliding

Foundations permitted to rock beyond $\frac{1}{2}$ uplift limit or exceed ultimate bearing stress and a linear stress distribution
OANR: Use $\frac{1}{2}$ uplift and linear stress distribution

More than the outer line of piles in group systems allowed to plunge or uplift under seismic loadings
OANR: Only outer line is permitted to reach tension capacity

Plumb piles that are not capacity-protected (e.g., integral abutment piles or pile-supported seat abutments that are not fused transversely)
OANR: Use seat abutment or a detail that allows movement

Batter pile systems in which the geotechnical capacities and/or in-ground hinging define the plastic mechanisms
OANR: Plastic hinging forced to occur above ground in column

Columns with architectural flares — with or without an isolation gap
Isolation gap optional
OANR: Remove flare

Note: OANR means a design alternate where owners approval is not required and a higher level of analysis (pushover in SDAP E) can be avoided.

FIGURE 4.9 Permissible earthquake resisting elements that require owner's approval [37].

curved superstructures, varying structure types, outriggers, unbalanced mass and stiffness, high skew) and unusual geologic conditions (soft soil, moderate to high liquefaction potential and proximity to an earthquake fault). In this case, project-specific criteria need to be developed to address their nonstandard features.

For important bridges such as the San Francisco-Oakland Bay Bridge and the Benicia-Martinez Bridge or structures in a designated lifeline route, performance-based project-specific two-level seismic design criteria [28–30] are required.

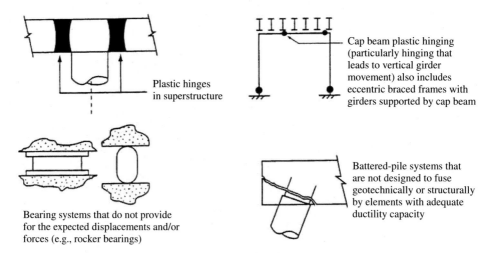

FIGURE 4.10 Earthquake resisting elements that are not recommended for new bridges (ATC/MCEER Joint Venture, 2001).

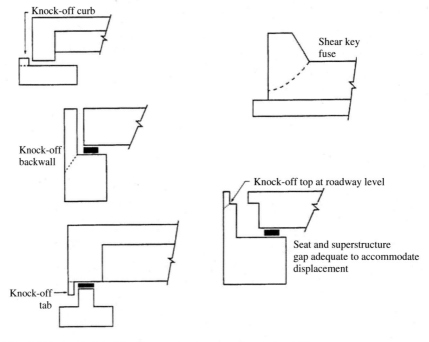

FIGURE 4.11 Methods of minimizing damage to abutment foundation [37].

The SDC shown in Table 4.2 shall be achieved by the following design objectives:

- All bridges shall be designed to withstand deformation imposed by the design earthquake.
- All structure components have sufficient strength and ductility to ensure collapse will not take place during the maximum credible earthquake (MCE). Ductile behavior can be provided by inelastic actions either through selected structural members and through protective systems — seismic isolations and energy dissipation devices.
- Inelastic behavior shall be limited to the preidentified locations, that is, ductile components explicitly designed for ductile performance, within the bridge that are easily inspected and

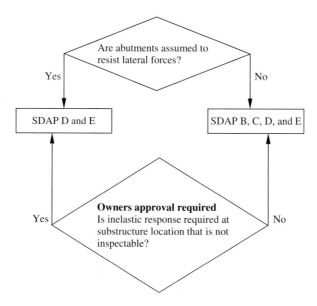

FIGURE 4.12 Design approaches for permissible earthquake resisting systems [37].

repaired following an earthquake. Because inelastic response of a concrete superstructure is difficult to inspect and repair and superstructure damage may cause a bridge to be unserviceable after an earthquake, inelastic behavior on most concrete bridges should be preferably be located in columns, pier walls, backwalls, and wingwalls.

- Structural components not explicitly designed for ductile performance, that is, capacity-protected components shall be designed to remain essentially elastic. That is, (1) response in concrete components shall be limited to minor cracking or limited to force demands not exceeding the nominal strength capacity determined by current Caltrans SDC [35] and (2) response in steel components shall be limited to force demands not exceeding the nominal strength capacity determined by the current Caltrans *Guide* [36]. To assure the yielding mechanism occurs in the desired location, the capacity design principle is used by providing overstrength for those capacity-protected components.

4.6 Seismic Design Approaches

4.6.1 ATC/MCEER Guidelines [37]

4.6.1.1 Seismic Loads

Seismic loads are represented by the design response spectrum curve for a damping ratio of 5% (Figure 4.13):

$$
S_a = \begin{cases} 0.6\dfrac{S_{DS}}{T_0}T + 0.4S_{DS} & \text{for } T \le T_0 = 0.2T_S \\[2mm] S_{DS} & \text{for } T_0 < T \le T_S = \dfrac{S_{D1}}{S_{DS}} \\[2mm] \dfrac{S_{D1}}{T} & \text{for } T > T_S \end{cases} \tag{4.1}
$$

$$
S_{DS} = F_a S_s \tag{4.2}
$$

$$
S_{D1} = F_v S_1 \tag{4.3}
$$

TABLE 4.1   Recommended LRFD Guidelines. Design Earthquakes and Seismic Performance Objectives [37]

Probability of exceedance for design earthquake ground motions[a]		Performance level[b]	
		Life safety	Operational
Rare earthquake (MCE)	Service[c]	Significant disruption	Immediate
3% PE in 75 years/1.5 mean deterministic	Damage[d]	Significant	Minimal
Expected earthquake 50% PE in 75 years	Service	Immediate	Immediate
	Damage	Minimal	Minimal to none

[a] The upper-level earthquake considered in the *Guidelines* is designated the Maximum Considered Earthquake or MCE. In general the ground motions or national MCE ground motion maps have a probability of exceedance (PE) of approximately 3% PE in 75 years. However, adjacent to highly active faults, ground motions on MCE maps are bounded deterministically as described in the Commentary for Article 3.2 [46]. When bounded deterministically, MCE ground motions have a probability of exceedance higher than 3% PE in 75 years not to exceed 1.5 times the mean deterministic values. The performance objective for the expected earthquake is either explicitly included as an elastic design for the 50% PE in 75-year force level or results implicitly from design for the 3% PE in 75-year force level.

[b] *Performance levels:* These are defined in terms of their anticipated performance objectives in the upper level earthquake. Life safety — In an MCE event, the bridge should not collapse but partial or complete replacement may be required. Since a dual level design is required, the life safety performance level will have immediate service and minimal damage for the expected design earthquake; Operational — For both rare and expected earthquakes, the bridge should be immediate service and minimal damage.

[c] *Service levels:* Immediate — Full access to normal traffic shall be available to traffic following an inspection of the bridge; Significant disruption — Limited access (reduced lanes, light emergency traffic) may be possible after shoring; however, the bridge may need to be replaced.

[d] *Damage levels:* None — Evidence of movement may be present but no notable damage; Minimal — Some visible signs of damage. Minor inelastic response may occur, but postearthquake damage is limited to narrow flexural cracking in concrete and the onset of yielding in steel. Permanent deformations are not apparent, and any repairs could be made under nonemergency conditions with the exception of superstructure joints; Significant — Although there is no collapse, permanent offsets may be occur and damage consisting of cracking, reinforcement yield, and major spalling of concrete and extensive yielding and local buckling of steel columns, global and local buckling of steel braces, and cracking in the bridge deck slabs at shear studs on the seismic load path is possible. These conditions may require closure to repair the damage. Partial or complete replacement of columns may be required in some cases. For sites with lateral flow due to liquefaction, significant inelastic deformation is permitted in the piles, whereas for all other sites the foundations are capacity-protected and no damage is permitted in the pile. Partial or complete replacement of the columns and piles may be necessary if significant lateral flow occurs. If replacement of columns or other components is to be avoided, the design approaches producing minimal or moderate damage such as seismic isolation or the control and repairability design concept should be assessed.

where S_{DS} is the design earthquake response spectral acceleration at short periods, S_{D1} is the design earthquake response spectral acceleration at 1-s period, S_s is the 0.2-s period spectral acceleration on Class B rock from national ground motion maps [45], S_1 is the 1-s period spectral acceleration on Class B rock from national ground motion maps [45], F_a is the site coefficient (Table 4.3) for the short-period portion of the design response spectrum curve, and F_v is the site coefficient (Table 4.4) for the long-period portion of the design response spectrum curve.

For Site Class F, which is not included in the Table 4.5, such as soils vulnerable to potential failure or collapse under seismic loading, such as liquefiable soils, quick and highly sensitive clays, and collapsible weakly cemented soils, site-specific geotechnical investigation and dynamic site response analyses shall be performed [37].

The effects of vertical ground motion may be ignored if the bridge site is located more than 50 km from an active fault. If the bridge is located within 10 km of an active fault, site-specific response spectra and acceleration time histories including vertical ground motions shall be considered.

TABLE 4.2 Caltrans Seismic Performance Criteria [24]

Ground motions at the site	Level of damage and postearthquake service	
	Ordinary bridge	Important bridge
Functional — evaluation ground motion	Service: immediate Damage: repairable	Service: immediate Damage: minimal
Safety — evaluation ground motion	Service: limited Damage: significant	Service: immediate Damage: repairable

Definition

Important bridge: A bridge meets one or more of the following requirements:
- Required to provide postearthquake life safety; such as access to emergency facilities.
- Time for restoration of functionality after closure would create a major economic impact.
- Formally designed as critical by a local emergency plan.

Ordinary bridge: Any bridge not classified as an important bridge.

Functional — evaluation ground motion (FEGM): This ground motion may be assessed either deterministically or probabilistically. The determination of this event is to be reviewed by a Caltrans-approved consensus group.

Safety — evaluation ground motion (SEGM): This ground motion may be assessed either deterministically or probabilistically. The deterministic assessment corresponds to the maximum credible earthquake (MCE). The probabilistic ground motion for the safety evaluation typically has a long return period (approximately 1000 to 2000 years).

MCE: The largest earthquake that is capable of occurring along an earthquake fault, based on current geologic information as defined in the 1996 Caltrans Seismic Hazard Map.

Service levels
- *Immediate:* Full access to normal traffic is available almost immediately following the earthquake.
- *Limited:* Limited access (e.g., reduced lanes, light emergency traffic) is possible with days of the earthquake. Full service is restorable within months.

Damage levels
- *Minimal:* Essentially elastic performance.
- *Repairable:* Damage that can be repaired with a minimum risk of losing functionality.
- *Significant:* A minimum risk of collapse, but damage that would require closure to repair.

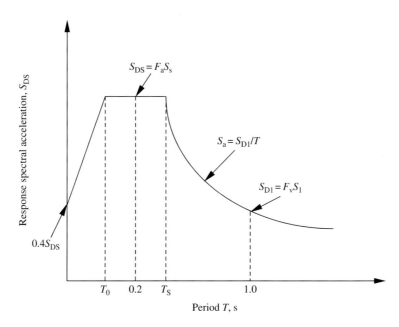

FIGURE 4.13 Design response spectrum curve [37].

TABLE 4.3 Recommended LRFD *Guidelines* — Site Coefficient F_a [37]

Site class	Mapped spectral response acceleration at short periods				
	$S_s \leq 0.25g$	$S_s = 0.5g$	$S_s = 0.75g$	$S_s = 1.00g$	$S_s = 1.25g$
A	0.8	0.8	0.8	0.8	0.8
B	1.0	1.0	1.0	1.0	1.0
C	1.2	1.2	1.1	1.0	1.0
D	1.6	1.4	1.2	1.1	1.0
E	2.5	1.7	1.2	0.9	0.9
F	a.0	a.0	a.0	a.0	a.0

[a] Site-specific geotechnical investigation and dynamic site response analysis shall be performed. For purpose of defining seismic hazard levels, Type E values may be used for Type F soils.

Note: Use straight line interpolation for intermediate values of S_s.

TABLE 4.4 Recommended LRFD *Guidelines* — Site Coefficient F_v [37]

Site class	Mapped spectral response acceleration at 1-s period				
	$S_s \leq 0.1g$	$S_s = 0.2g$	$S_s = 0.3g$	$S_s = 0.4g$	$S_s \geq 0.5g$
A	0.8	0.8	0.8	0.8	0.8
B	1.0	1.0	1.0	1.0	1.0
C	1.7	1.6	1.5	1.4	1.3
D	2.4	2.0	1.8	1.6	1.5
E	3.5	3.2	2.8	2.4	2.4
F	a.0	a.0	a.0	a.0	a.0

[a] Site-specific geotechnical investigation and dynamic site response analysis shall be performed. For purpose of defining seismic hazard levels, Type E values may be used for Type F soils.

Note: Use straight line interpolation for intermediate values of S_1.

TABLE 4.5 Recommended LRFD *Guidelines* — Site Classification [37]

Site class	$\bar{v}_s$ (m/s)	$\bar{N}$ or $\bar{N}_{ch}$ (blows/0.3 m)	$\bar{s}_u$ (kPa)
E	<180	<15	<50
D	180 to 360	15 to 50	50 to 100
C	360 to 760	>50	>100
B	760 to 1500	—	—
A	>1500	—	—

Notes:

$\bar{v}_s$ is the average shear wave velocity for top 30 m of a site.

$\bar{N}$ is the average standard penetration test blow count for top 30 m of a site.

$\bar{N}_{ch}$ is the average standard penetration test blow count for cohesionless layers of top 30 m of a site.

$\bar{s}_u$ is the average undrained shear strength of cohesive layers for top 30 m of a site.

Seismic force effects from different vibration modes shall be combined. For modal response closely spaced in frequency and ground motion in one direction, the complete quadratic combination (CQC) method shall be used. For combining the contribution of orthogonal and uncorrelated ground motion components to a single seismic load case, the square root sum of the squares (SSRC) method or 100 to 40% rule shall be used.

For a bending moment, the combination rules are as follows:

SSRC combination:

$$M_x = \sqrt{\left(M_x^T\right)^2 + \left(M_x^L\right)^2 + \left(M_x^V\right)^2} \tag{4.4}$$

100 to 40% combination:

$$M_x^{\text{LC1}} = 1.0M_x^T + 0.4M_x^L + 0.4M_x^V \tag{4.5}$$

$$M_x^{\text{LC2}} = 0.4M_x^T + 1.0M_x^L + 0.4M_x^V \tag{4.6}$$

$$M_x^{\text{LC3}} = 0.4M_x^T + 0.4M_x^L + 1.0M_x^V \tag{4.7}$$

For circular columns, the vector moments and axial forces shall be obtained for biaxial design:

SSRC combination:

- For bridges with skew angle less than $10°$, the maximum of $\sqrt{M_x^2 + (0.4M_y)^2}$ and $\sqrt{M_y^2 + (0.4M_x)^2}$ with the maximum axial load $\pm P$
- For bridges with skew angle greater than $10°$, $\sqrt{M_x^2 + M_y^2}$ with the maximum axial load $\pm P$

100 to 40% combination:

- The maximum of $\sqrt{(M_x^{\text{LC1}})^2 + (M_y^{\text{LC1}})^2}$ and $\sqrt{(M_x^{\text{LC2}})^2 + (M_y^{\text{LC2}})^2}$ and $\sqrt{(M_x^{\text{LC3}})^2 + (M_y^{\text{LC3}})^2}$ with the maximum axial load $\pm P$

where subscripts x and y represent two horizontal axes, x–x and y–y, respectively; superscripts L, T, and V indicate the longitudinal, transverse, and vertical directions, respectively; and superscripts LC1, LC2, and LC3 are load cases 1, 2, and 3, respectively.

4.6.1.2 Seismic Design and Analysis Procedures (SDAP)

Depending on the seismic hazard levels specified in Table 4.6, each bridge shall be designed, analyzed, and detailed in accordance with Table 4.6.

4.6.1.2.1 Single-Span Bridges

Single-span bridges need not be analyzed for seismic loads and design requirements are limited to the minimum seat widths and connection forces, which shall not be less than the product of $F_a S_S/2.5$ and the tributary permanent load.

4.6.1.2.2 SDAP A1 and A2

For low seismicity areas, only minimum seat widths and connection design forces for bearings and minimum shear reinforcement in concrete columns and piles in the seismic design requirement (SDR) 2 are deemed necessary for the life safety performance objective. The primary purpose is to ensure that the connections between the superstructure and its supporting substructures remain intact during the design earthquake. SDAP A1 and A2 require that the horizontal design connection forces in the restrained directions shall not be taken to be less than 0.1 and 0.25 times the vertical reactions due to tributary permanent loads and assumed existing live loads, respectively.

TABLE 4.6 Recommended LRFD *Guidelines* — SHL, SDAP, and SDR [37]

Seismic hazard level (SHL)	S_{D1} ($F_v S_1$)	S_{DS} ($F_a S_s$)	Seismic design and analysis procedure (SDAP) and seismic design requirements (SDR)			
			Life safety		Operational	
			SDAP	SDR	SDAP	SDR
I	$0.15 < S_{D1} \leq 0.15$	$0.15 < S_{DS} \leq 0.15$	A1	1	A2	2
II	$0.15 < S_{D1} \leq 0.25$	$0.15 < S_{DS} \leq 0.35$	A2	2	C/D/E	3
III	$0.25 < S_{D1} \leq 0.40$	$0.35 < S_{DS} \leq 0.60$	B/C/D/E	3	C/D/E	5
IV	$0.40 < S_{D1}$	$0.60 < S_{DS}$	C/D/E	4	C/D/E	6

4.6.1.2.3 SDAP B — No-Analysis Approach

The no-analysis approach allows for the bridge to be designed for all nonseismic requirements without a seismic demand analysis and the capacity design principle is used for all components connected to columns. For geotechnical design of the foundations, the moment overstrength capacity of columns that frame into the foundation, $M_{po} = 1.0M_n$, where M_n is nominal moment capacity of a column. SDAP B applies only to regular bridges meeting the following restrictions:

- The maximum span length is less than both 80 m and 1.5 times the average span length.
- The maximum skew angle is less than 30°.
- The ratio of the maximum interior bent stiffness to the average bent stiffness of the bridge is less than 2.
- The subtended angle in the horizontally curved bridges is less than 30°.
- For frames in which the superstructure is continuous over the bents and some bents do not participate in the ERS, $F_v S_1 (N_{bent}/N_{ers}) < 0.4 \cos \alpha_{skew}$, where N_{bent} and N_{ers} are total number of bents in the frame and number of bents participating in the ERS in the longitudinal direction, respectively; and α_{skew} is the skew angle of the bridge.
- $F_v S_1 < 0.4 \cos \alpha_{skew}$.
- The bridge site has a low potential for liquefaction and the piers are not seated on spread footing.
- For concrete column and pile bents, column axial load, $P_e < 0.15 f'_c A_g$ (f'_c is specified minimum concrete compression strength and A_g is gross cross-sectional area of column); longitudinal reinforcement ratio, $\rho_1 > 0.008$; column transverse dimension, $D > 300$ mm; and maximum column moment–shear ratio, $M/VD < 6$.
- For concrete wall piers with low volumes of longitudinal steel, $P_e < 0.1 f'_c A_g$; $\rho_1 > 0.0025$; wall thickness or smallest cross-sectional dimension, $t > 300$ mm; and $M/Vt < 10$.
- For steel pile bents framing into concrete caps, $P_e < 0.15 P_y$ (P_y is axial yield force of steel pile); pile dimension about the weak axis bending at the ground level, $D_p > 250$ mm; and $L/b < 10$ (L is the length from point of maximum moment to the inflexion point of the pile when subjected to a pure transverse load and b is the flange width; for a cantilever column in a pile bent configuration, L is equal to the length above ground to the top of the bent plus 3 pile diameters).
- For timber piles framing into concrete caps or steel moment–frame columns, $P_e < 0.1 P_c$ (and P_c is axial compression capacity of the pile or the column); $D_p > 250$ mm; and $M/VD_p < 10$.

4.6.1.2.4 SDAP C — Capacity Spectrum Design Method

The capacity spectrum design method combines a demand and capacity analysis and is conceptually the same as the Caltrans displacement-based design method. The primary difference is that the SDAP C begins with nonseismic design and then assesses the adequacy of the displacement. The key equations is the relationship between the seismic coefficient C_s and displacement, Δ:

$$C_s \Delta = \left(\frac{F_v S_1}{2 \pi B_L} \right)^2 g \tag{4.8}$$

$$C_s = \frac{F_a S_s}{B_S} \tag{4.9}$$

where B_L is the response reduction factor for long period structures as specified in Table 4.7, B_S is the response reduction factor for short period structures as specified in Table 4.7, g is the acceleration due to gravity (9.8 m/s²), and Δ is the lateral displacement of the pier; taken as 1.3 times the yield displacement of the pier when the long period equation governs.

The lesser of Equations 4.8 and 4.9 shall be used to assess C_s for two-level earthquakes. The required lateral strength of the bridge is $V_n = C_s W$ where W is the weight of the bridge responding

TABLE 4.7 Recommended LRFD *Guidelines* — Capacity Spectrum Response Reduction Factors for Bridge with Ductile Piers [37]

Earthquake	Performance level	B_s	B_L
50% PE in 75-year	Life safety	1	1
	Operational	1	1
3% PE in 75-year/1.5 mean deterministic	Life safety	2.3	1.6
	Operational	1	1

to earthquake ground motion. The procedure applies only to bridges that behave essentially as a single degree-of-freedom system and very regular bridges satisfying the following requirements:

- The number of spans per frame does not exceed six.
- The number of spans per frame is at least three, unless seismic isolation bearings are utilized at the abutments.
- The maximum span length is less than both 60 m and 1.5 times the average span length in a frame.
- The maximum skew angle is less than 30° and skew of piers or bents differs by less than 5° in the same direction.
- The subtended angle in the horizontally curved bridges is less than 20°.
- The ratio of the maximum bent or pier stiffness to the average bent stiffness is less than 2, including the effects of foundation.
- The ratio of the maximum lateral strength (or seismic coefficient) to the average bent strength is less than 1.5.
- Abutment shall not be assumed to resist the significant forces in both the transverse and longitudinal directions. Pier wall substructures must have bearings to permit transverse movement.
- For concrete column and pile bents, $P_e \leq 0.2 f'_c A_g$; $\rho_1 > 0.008$; and $D > 300$ mm.
- Piers and bents must have pile foundations when the bridge site has a potential for liquefaction.

4.6.1.2.5 SDAP D — Elastic Response Spectrum Method

The elastic response spectrum method uses either the uniform load or multimode method of analysis by considering cracked section properties. The analysis shall be performed for the governing design earthquakes, either the 50% PE (probability of exceedence) in 75-year or the 3% PE in 75-year/1.5 mean deterministic earthquake. Elastic forces obtained from analyses shall be modified using the response modification factor *R*.

4.6.1.2.6 SDAP E — Elastic Response Spectrum Method with Displacement Capacity Verification

SDAP E is a two-step design procedure. The first step is the same as SDAP D and the second step is to perform a two-dimensional nonlinear static (push over) analysis to verify substructure displacement capacity.

4.6.1.3 Response Modification Factor *R*

It is generally recognized that it is uneconomical, sometimes impractical, and impossible to design a bridge to resist large earthquake elastically. Columns are assumed to deform inelastically where seismic forces exceed their design levels, which is established by dividing the elastically computed force effects by the response modification factor *R*. The *R*-factors specified in the following were based on an evaluation of existing test data, engineering judgement, and the equal displacement principle as shown in Figure 4.14. It is used in principle for the conventional ductile design.

For substructures

$$R = 1 + (R_B - 1)\frac{T}{T^*} \leq R_B \qquad (4.10)$$

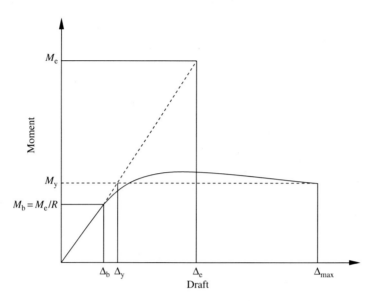

FIGURE 4.14 Basis for conventional ductile design [37].

TABLE 4.8 Recommended LRFD *Guidelines* — Base Response Modification Factor for Substructures, R_B (ATC/MCEER Joint Venture, 2001)

	Performance objective			
	Life safety		Operational	
Substructure element	SDAP D	SDAP E	SDAP D	SDAP E
Wall-type pier — large dimension	2.0	3.0	1.0	1.5
Columns — single and multiple	4.0	6.0	1.5	2.5
Pile bents and drilled shafts — vertical piles — above ground	4.0	6.0	1.5	2.5
Pile bents and drilled shafts — vertical piles — 2 diameters below ground level — no owners approval required	1.0	1.5	1.0	1.0
Pile bents and drilled shafts — vertical piles — in ground — owners approval required	N/A	2.5	N/A	1.5
Pile bents with batter piles	N/A	2.5	N/A	1.5
Seismically isolated structures	1.5	1.5	1.0	1.5
Steel-braced frame — ductile components	3.0	4.5	1.0	1.5
Steel-braced frame — normally ductile components	1.5	2.0	1.0	1.0
All elements for expected earthquake	1.3	1.3	0.9	0.9

Notes:

1. The substructure design forces resulting from the elastic analysis divided by the appropriate R-factor for SDAP E can not be reduced below 70% at the R-factored reduced forces or the 50% PE in 75-year design forces as part of the pushover analysis.
2. There may be design situation (e.g., architecturally oversized column) where a designer opts to design the column for an $R = 1$ (i.e., elastic design). In concrete columns the associate elastic design shear may be obtained from the elastic analysis forces using an R-factor of 0.67 or by calculating the design shear by the capacity design procedures using a flexural overstrength factor of 1.0. In steel-braced frame if an $R = 1.0$ is used the connection design forces shall be obtained using an $R = 0.67$. If an $R = 1.0$ is used in any design the foundation shall be designed for elastic forces plus the SDR detailing requirements are required for concrete piles (i.e., minimum shear requirements).
3. Unless specifically stated, the R-factor applies to both steel and concrete.
4. N/A means that owners approval is required and thus the use of SDAP E is required.

where R_B is the base response modification factor specified in Table 4.8, T is the natural period of the structure, $T^* = 1.25T_s$ where T_s is as defined in Section 4.6.1.1.

For connections (superstructure to abutment; expansion joints within a span of the superstructure, columns, piers, or pile bents to cap beam or superstructure; and column or piers to foundations), an R-factor of 0.8 shall be used for those cases where capacity design principles are not used to develop the design forces to design the connections. It is assumed that if the $R < 1.5$, columns should remain essentially elastic for design earthquake; if $1.5 < R > 3.0$, columns should be repairable; if $R > 3.0$ significant plastic hinging may occur and the column may not be repairable; however, collapse is still prevented.

4.6.1.4 Capacity Design Principle

The main objective of the capacity design principle is to ensure the desirable mechanisms can dissipate significant amounts of energy and inelastic deformation (plastic hinging) occurs at expected locations (at top and bottom of columns) where they can be readily inspected and repaired. To achieve this objective, the overstrength force effects developed from the plastic hinges in columns shall be dependably resisted by column shear and adjoining elements such as cap beams, spread footing, pile cap, and foundations. The moment overstrength capacity (M_{po}) can be assessed using one of the following approaches:

- $M_{po} = \begin{cases} 1.5M_n & \text{for concrete column} \\ 1.2M_n & \text{for steel column}, M_n \text{ based on expected yield strength} \\ 1.3M_n & \text{for concrete filled steel tubes} \\ 1.5M_n & \text{for steel piles in weak axis bending and for steel members} \\ & \quad \text{in shear (e.g., eccentrically braced frames)} \\ 1.0M_n & \text{for geotechnical design force in SDR3} \end{cases}$ (4.11)

- For reinforce concrete columns [46]

$$M_{po} = M_{bo}\left[1 - \left(\frac{P_e - P_b}{P_{to} - P_b}\right)^2\right]$$ (4.12)

where

P_e = axial compression load based on gravity load and seismic (framing) action
P_b = axial compression capacity at the maximum nominal (balanced) moment on the section
   $= 0.425\,\beta_1 f_c' A_g$
   β_1 = compression stress block factor ≤ 0.85
P_{to} = axial tensile capacity of the column $= -A_{st}f_{su}$
   A_{st} = area of longitudinal reinforcement
   f_{su} = ultimate tensile strength of the longitudinal reinforcement

$$M_{bo} = K_{shape}A_{st}f_{su}D' + P_bD\left(\frac{1 - \kappa_0}{2}\right)$$ (4.13)

D' = pitch circular diameter of the reinforcement in a circular section or the out-to-out dimension of the reinforcement in a rectangular section, this generally may be assumed as $= 0.8D$

$$K_{shape} = \begin{cases} 0.32 & \text{for circular sections} \\ 0.375 & \text{for square sections with 25\% of the longitudinal} \\ & \quad \text{reinforcement placed in each face} \\ 0.25 & \text{for walls with strong axis bending} \\ 0.5 & \text{for walls with weak axis bending} \end{cases}$$ (4.14)

κ_0 = a factor related to the centroid of compression stress block and should be taken as 0.6 and 0.5 for circular and rectangular sections, respectively.

It should be pointed out that Equations 4.12 and 4.13 are rearranged in the simpler format given above.

- For reinforced concrete column, a moment–curvature section analysis taking into account the expected strength, confined concrete properties, and strain hardening effects of longitudinal reinforcement.
- For a steel column, nominal flexural resistance (M_n) shall be determined either in accordance with the AASHTO-LRFD [38] or

$$M_n = 1.18 M_{px}\left[1 - \frac{P_u}{A_g F_{ye}}\right] \le M_{px} \qquad (4.15)$$

A_g = gross cross-sectional area of a steel column
F_{ye} = expected specified minimum yield strength of steel
M_{px} = plastic moment under pure bending calculated using F_{ye}
P_u = factored axial compression load

4.6.1.5 Plastic Hinge Zones

The plastic hinge zones (L_p) for typical concrete and steel columns, pile bents, and drilled shaft and zones of a columns above a footing or above an oversized in-ground drilled shaft shall be the maximum of the following:

For reinforced concrete columns

$$L_p = \text{maximum of} \begin{cases} D_{\max} \\ \dfrac{L}{6} \\ D\left(\cot\theta + \dfrac{\tan\theta}{2}\right) \\ 1.5\left(0.08\dfrac{M}{V} + 4000\,\varepsilon_y d_b\right) \\ \dfrac{M}{V}\left(1 - \dfrac{M_y}{M_{po}}\right) \\ 450\,\text{mm} \end{cases} \qquad (4.16)$$

where

D = transverse column dimension in the direction of bending
$D_{\max}$ = maximum cross-sectional dimension of a column
d_b = diameter of longitudinal reinforcement
L = clear height of a column
M = maximum column moment
V = maximum column shear
M_y = column yield moment
ε_y = yield strain of longitudinal reinforcement
θ = principal crack angle
= $\tan^{-1}((1.6/\Lambda)(\rho_v/\rho_t)(A_v/A_g))^{0.25}$ with $\theta \ge 25°$ and $\theta \ge \tan^{-1}(D'/L)$
  A_v = shear area of concrete which may be taken as $0.8A_g$ and $b_w d$ for a circular section and a rectangular section, respectively
  ρ_v = ratio of transverse reinforcement
  ρ_t = volumetric ratio of longitudinal reinforcement
  Λ = fixity factor taken as 1 for fixed-pinned and 2 for fixed-fixed ends

For steel columns:

$$L_p = \text{maximum of} \begin{cases} \dfrac{L}{8} \\ 450\,\text{mm} \end{cases} \qquad (4.17)$$

For a flared column, the plastic hinge zone shall be extended from the top of column to a distance equal to the maximum of the above criteria below the bottom of the flare. The areas within the plastic hinge zones shall be detailed to assure ductile element behavior (e.g., providing confinement reinforcement in concrete columns or inelastic local buckling behavior in steel columns).

4.6.2 Caltrans Seismic Design Criteria

4.6.2.1 Seismic Loads

For ordinary bridges, safety-evaluation ground motion is based on deterministic assessment corresponding to the MCE, the largest earthquake, which is capable of occurring based on current geologic information. A set of ARS curves developed by ATC-32 are adopted as standard horizontal ARS curves in conjunction with the peak rock acceleration from the Caltrans Seismic Hazard Map 1996 to determine the horizontal earthquake forces. Figure 4.15 shows typical ARS curves. Vertical acceleration shall be considered for bridges with nonstandard structural components, unusual site conditions, and/or close

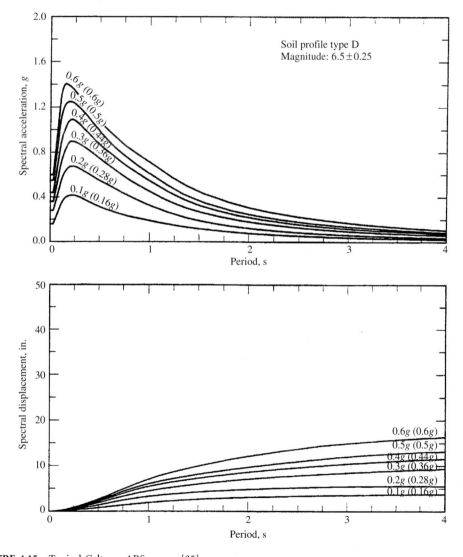

FIGURE 4.15 Typical Caltrans ARS curves [35].
Note: Peak ground acceleration values not in parentheses are for rock (soil profile type B) and peak ground acceleration values in parentheses are for soil profile type D.

proximity to earthquake faults and can be approximated by an equivalent static vertical force applied to the superstructure.

For structures within 15 km from an active fault, the spectral ordinates of the appropriate standard ARS curves shall be increased by 20%. For long period structures ($T \geq 1.5$ s) on deep soil sites (depth of alluvium ≥ 75 m) the spectral ordinates of the appropriate standard ARS curves shall be increased by 20% and the increase applies to the portion of the curves with periods greater than 1.5 s.

4.6.2.2 Design Approaches

The displacement-based design approach is used to ensure that the structural system and its individual components have enough displacement capacities to withstand the deformation imposed by the design earthquake. Using displacements rather than forces as a measurement of earthquake damages allows a structure to fulfill the required functions.

The displacements of the global system and the local ductile system shall satisfy the following requirement:

$$\Delta_D \leq \Delta_C \qquad (4.18)$$

where Δ_D is the displacement demand determined by the global analysis, the stand-alone analysis, or the larger of the two if both types of analyses are necessary and Δ_C is the displacement capacity when any plastic hinge capacity reaches its ultimate capacity.

In a displacement-based analysis, proportioning of the structure is made first based on strength and stiffness requirements. The appropriate seismic analysis is performed and the resulting displacements are compared to the displacement capacity, which is dependent on the formation and rotational capacity of plastic hinges and can be evaluated by an inelastic static pushover analysis. This procedure has been used widely in seismic bridge design in California since 1994. Figure 4.16 shows the design flowchart for the ordinary standard bridges.

4.6.2.3 Displacement Demands on Bridges and Ductile Components

Seismic demands on bridge systems and ductile components are measured in terms of displacements rather than forces. Displacement demands shall be estimated from either equivalent static analysis (ESA) or elastic dynamic analysis (EDA, i.e., elastic response spectrum analysis) for typical bridge periods of 0.7 to 3 s. Attempts should be made to design bridges with dynamic characteristics (mass and stiffness) so that the fundamental period falls within the region between 0.7 and 3 s where the equal displacement principle applies. For short period bridges, linear elastic analysis typically underestimates displacement demands. In these cases, the inability to accurately predict displacements can be overcome by either designing the bridge to perform elastically, multiplying the elastic displacement from analysis by an amplification factor, or using seismic isolation and energy dissipation devices to limit seismic response. For long period ($T > 3$ s) bridges, a linear elastic analysis generally overestimates displacements and a linear elastic displacement response spectrum analysis should be used.

4.6.2.4 Force Demands on Capacity-Protected Components

Seismic demands on capacity-protected components, such as superstructures, bent caps, and foundations, are measured in terms of forces rather than displacements. Force demands for capacity-protected components shall be determined by the joint-force equilibrium considering plastic hinging capacity of the ductile component multiplied by an overstrength factor of 1.2. The overstrength factor accounts for the possibility that the actual ultimate plastic moment strength of the ductile component exceeds its estimated idealized plastic capacity based on the expected properties. This overstrength factor (1.2) is utilized not only to determine the demands on the capacity-protected members, but is also used to determine shear demands on the columns themselves.

4.6.2.5 Seismic Capacity

4.6.2.5.1 General

Strength and displacement capacities of a ductile flexural element shall be evaluated by moment–curvature analysis based on the expected material properties and anticipated damages. The impact the second-order

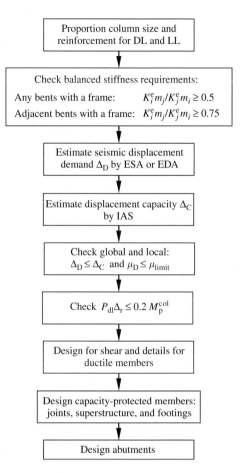

FIGURE 4.16 Caltrans seismic procedure for ordinary standard bridges.

P–Δ effect on the strength and displacement capacities of all members subjected to combined bending and compression shall be considered. Components may require redesign if the P–Δ effect is significant.

4.6.2.5.2 Displacement Capacity

The displacement capacity of a bridge system shall be evaluated by an inelastic static analysis (i.e., a static push over analysis). The rotational capacity of all plastic hinges shall be limited to a safe performance level. The plastic hinge regions shall be designed and detailed to perform with minimal strength degradation under cyclic loading.

The displacement capacity of a local member can be evaluated by its rotational capacity. The displacement capacity of a prismatic cantilever member (Figure 4.17) can be calculated as

$$\Delta_c = \Delta_Y^{col} + \Delta_p \tag{4.19}$$

$$\Delta_Y^{col} = L^2/3 \times \phi_Y \tag{4.20}$$

$$\Delta_p = \theta_p \times \left(L - \frac{L_p}{2} \right) \tag{4.21}$$

$$\theta_p = L_p \times \phi_p \tag{4.22}$$

$$\phi_p = \phi_u - \phi_Y \tag{4.23}$$

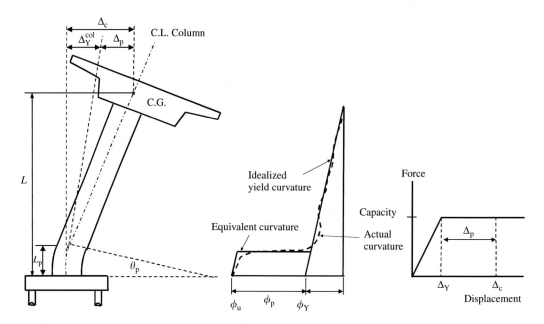

FIGURE 4.17   Displacement of a cantilever member [35].

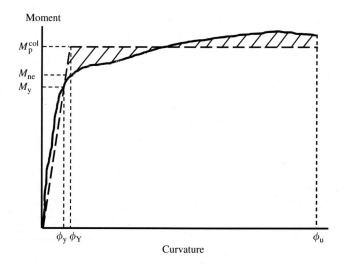

FIGURE 4.18   Idealized moment–curvature curve [35].

where

L = distance from the point of maximum moment to the point of contraflexure

L_p = equivalent analytical plastic hinge length as defined in Section 4.8.2.6

Δ_p = idealized plastic displacement capacity due to rotation of the plastic hinge

Δ_Y^{col} = idealized yield displacement of the column at the formation of the plastic hinge

ϕ_Y = idealized yield curvature defined by an elastic–perfectly plastic representation of the cross-section's M–ϕ curve, see Figure 4.18

ϕ_p = idealized plastic curvature capacity (assumed constant over L_p)

ϕ_u = curvature capacity at the failure limit state, defined as the concrete strain reaching ε_{cu} or the main column reinforcing steel reaching the reduced ultimate strain ε_{su}^R

θ_p = plastic rotation capacity of an equivalent plastic hinge

However, it should be pointed out that Equation 4.19 might overestimate the displacement capacity for a reinforced concrete column [47]. Column slenderness, high compression axial loads, and a low percentage of reinforcement all may contribute to the overestimating of the displacement capacity. Special attention, therefore, should be paid to the estimation of displacement capacity. It was recommended [47] that the P–Δ effect should be taken into account in calculating lateral load-carrying capacity and the displacement capacity, especially for medium-long and long columns. The lateral displacement capacity Δ_c can be chosen as the displacement that corresponds to the condition when lateral load carrying capacity degrades to a certain acceptable level, say a minimum of 80% of the peak resistance (Figure 4.19) or peak load [47–49].

4.6.2.5.3 Shear Capacity

Shear capacity of concrete members shall be calculated using nominal material properties as

$$\phi V_n = \phi(V_c + V_s) \tag{4.24}$$

$$\phi = 0.85 \tag{4.25}$$

$$V_c = v_c(0.8 A_g) \tag{4.26}$$

Concrete shear capacity is influenced by flexural and axial loads and is calculated separately for regions within the plastic hinge zone and regions outside this zone. In the plastic hinge zone, concrete shear capacity is modified based on the level of confinement and the displacement ductility demand (Figure 4.20).

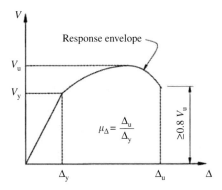

FIGURE 4.19 Lateral load–displacement curve [36].

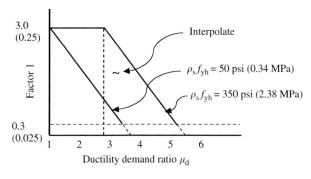

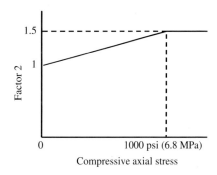

FIGURE 4.20 Shear factors [35].

- Inside the plastic hinge zone

$$v_c = \text{Factor } 1 \times \text{Factor } 2 \times \sqrt{f_c'} \leq 0.33\sqrt{f_c'} \quad \text{(MPa)} \tag{4.27}$$

- Outside the plastic hinge zone

$$v_c = 0.25 \times \text{Factor } 2 \times \sqrt{f_c'} \leq 0.33\sqrt{f_c'} \quad \text{(MPa)} \tag{4.28}$$

$$\text{Factor } 1 = 0.025 \leq \frac{\rho_s f_{yh}}{12.5} + 0.305 - 0.083\mu_d < 0.25 \tag{4.29}$$

$$\text{Factor } 2 = 1 + \frac{P_c}{13.8\, A_g} \leq 1.5 \tag{4.30}$$

To ensure reliable capacity in the plastic hinge regions, all column lateral reinforcements are required to be butt welded or spliced hoops capable of resisting the ultimate capacity of the reinforcing steel.

- For confined circular or interlocking core sections:

$$V_s = \frac{\pi}{2}\frac{n A_b f_{yh} D'}{s} \tag{4.31}$$

- For pier wall in weak direction

$$V_s = \frac{A_v f_{yh} d}{s} \tag{4.32}$$

where A_b is the area of an individual interlock spiral or hoop bar, A_v is the total area of shear reinforcement perpendicular to flexural tension reinforcement, D' is the cross-section dimension of confined concrete core measured between the centerline of the peripheral hoop or spiral bars, f_{vh} is the specified minimum yield strength of transverse reinforcement, d is the area of shear reinforcement perpendicular to flexural tension reinforcement, n is the number of individual interlock spirals or hoops core section, and s is the spacing of transverse reinforcement.

4.7　Seismic Analysis and Modeling

4.7.1　Equivalent Static Analysis

The ESA method specified in Caltrans SDC [35] can be used for simpler structures, which have balanced spans and similar bent stiffness. Low skew and seismic response are primarily captured by the fundamental mode of vibration. In the ESA method, the fundamental period is determined using tributary mass and stiffness at each bent. The applied seismic force is the product of the period dependent ARS coefficient and the tributary weight.

4.7.2　Elastic Response Spectrum Analysis (ERSA)

The ERSA including uniform load method and multimode dynamic analysis method is a linear elastic spectral analysis with the appropriate response spectrum. An adequate number of modes considered to capture a minimum of 90% mass participation shall be used for complex structures.

The uniform load method specified in ATC/MCEER *Guidelines* [37] is essentially an ESA method that uses a uniform lateral load distribution to approximate the effect of seismic loads. It may be used for both transverse and longitudinal directions if structures satisfy the requirements in Table 4.9.

TABLE 4.9 Recommended LRFD *Guidelines* — Requirement for Uniform Load Method [37]

Parameter	Value				
Number of spans	2	3	4	5	6
Maximum subtended angle for a horizontally curved bridge	20°	20°	30°	30°	30°
Maximum span length ratio from span to span	3	2	2	1.5	1.5
Maximum bent/pier stiffness ratio from span to span, excluding abutments	—	4	4	4	2

In both ESA and ERSA analyses, "effective" stiffness of the components shall be used in order to obtain realistic evaluation for the structure's period and displacement demands. The effective stiffness of ductile components shall represent the component's actual secant stiffness near first yield of rebar. The effective stiffness shall include the effects of concrete cracking, reinforcement, and axial load for concrete components; residual stresses, out-of-straightness, and axial load for steel components; and the restraints of the surrounding soil for piles. For ductile concrete column members, effective moments of inertia, I_{eff}, shall be based on cracked section properties and can be determined from the initial slope of the M–ϕ curve between the origin and the point designating first yield of the main column reinforcement. The torsional moment of inertia of concrete column J_{eff} may be taken as 0.2 times J_{gross}. For capacity-protected concrete members, I_{eff} shall be based on their level of cracking. For a conventionally reinforced concrete box girder superstructure, I_{eff} can be estimated between 0.5 and 0.75 times I_{gross}, the moment of inertia of a gross section. For prestressed concrete super-structures, I_{eff} is assumed 1.0 times I_{gross}.

The following are major considerations in seismic analysis and design practice:

- A beam–element model with three or more lumped masses for each member is usually used [25,26].
- Larger cap stiffness is often used to simulate a stiff deck.
- Compression and tension models are used to simulate the behavior of expansion joints. In the tension model, superstructure joints including abutments are released longitudinally but the restrainers are modeled as truss elements. In the compression model, all restrainers are considered to be inactive and all joints are locked longitudinally.
- Simplistic analysis models should be used for initial assessment of structural behavior. The results of more sophisticated models shall be compared for reasonableness with the results obtained from the simplistic models. The rotational and translational stiffness of abutments and foundations modeled in the seismic analysis must be compatible with their structural and geotechnical capacity. The energy dissipation capacity of the abutments should be considered for bridges whose response is dominated by the abutments [50].
- For elastic response spectrum analysis, the viscous damping ratio inherent in the specified ground spectra is usually 5%.
- For time history analysis, in lieu of measurements, a damping ratio of 5% for both concrete and timber constructions and 2% for welded and bolted steel construction may be used.
- For one- or two-span continuous bridges with abutment designed to activate significant passive pressure in the longitudinal direction, a damping ratio of up to 10% may be used in longitudinal analysis.
- Soil-spring elements should be used to the soil–foundation–structure interaction. Adjustments are often made to meet force–displacement compatibility, particularly for abutments. The maximum capacity of the soil behind abutments with heights larger than 2.5 m may be taken as 370 kPa and will be linearly reduced for the backwall height less than 2.5 m.

- Pile footing with pile cap and spread footing with soil types A and B [37] may be modeled as rigid. If footing flexibility contributes more than 20% to pier displacement, foundation springs shall be considered.
- For pile bent/drilled shaft, estimated depth to fixity or soil-spring based on idealized p–y curves should be used.
- Force–deformation behavior of a seismic isolator can be idealized as a bilinear relationship with two key variables: second slope stiffness and characteristic strength. For design, the force–deformation relationship can be represented by an effective stiffness based on the secant stiffness and a damping coefficient. For more detailed information, references can be made ATC/MCEER *Guidelines* and AASHTO Guide Specifications [37,51] and a comprehensive chapter by Zhang [52].

4.7.3 Nonlinear Dynamic Analysis

The nonlinear dynamic analysis (NDA) procedure is normally used for the 3% PE in 75-year earthquake. A minimum of three ground motions including two horizontal components and one vertical component shall be used and the maximum actions for those three motions shall be used for design. If more than seven ground motions are used, the design action may be taken as the mean action of ground motions. The result of an NDA should be compared with an ERSA as a check for a reasonableness of the nonlinear model.

4.7.4 Global and Stand-Alone Analysis

The global analysis specified in Caltrans SDC [35] is an EDA considering the entire bridge modeled from abutment to abutment. It is often used to determine displacement demands on multiframe structures. The stand-alone analysis is an elastic dynamic analysis considering only one individual frame. To avoid having individual frames dependent on the strength and stiffness of adjacent frames, the separate stand-alone model for each frame must meet all requirements of the SDC.

4.7.5 Inelastic Static Analysis — Push Over Analysis

Inelastic Static Analysis (ISA), commonly referred to as the "push over analysis," shall be used to determine the displacement capacity of a bridge system. IAS shall be performed using expected material properties for modeled members. ISA can be categorized into three types of analysis: (1) elastic–plastic hinge, (2) refined plastic hinge, and (3) distributed plasticity.

The simplest method, elastic–plastic hinge analysis, may be used to obtain an upper bound solution. The most accurate method, distributed plasticity analysis, can be used to obtain a better (more refined) solution. Refined plastic hinge analysis is an alternative that can reasonably achieve both computational efficiency and accuracy.

In an elastic–plastic hinge (lumped plasticity) analysis, material inelasticity is taken into account using concentrated "zero-length" plastic hinges, which maintain plastic moment capacities and rotate freely. When the section reaches its plastic capacity, a plastic hinge is formed and element stiffness is adjusted [53,54]. For regions in a framed member away from the plastic hinge, elastic behavior is assumed. It does not, however, accurately represent the distributed plasticity and associated P–δ effects. This analysis predicts an upper bound solution.

In the refined plastic hinge analysis [55], a two-surface yield model considers the reduction of plastic moment capacity at the plastic hinge due to the presence of axial force and an effective tangent modulus accounts for the stiffness degradation due to distributed plasticity along a frame member. This analysis is similar to the elastic–plastic hinge analysis in efficiency and simplicity and also accounts for distributed plasticity.

Distributed plasticity analysis models the spread of inelasticity through the cross-sections and along the length of the members. This is also referred to as plastic zone analysis, spread-of-plasticity analysis, and elasto-plastic analysis by various researchers. In this analysis, a member needs to be subdivided into several elements along its length to model the inelastic behavior more accurately. Two main approaches have been successfully used to model plastification of members in a second-order distributed plasticity analysis:

- Cross-sectional behavior is described as an input for the analysis by means of moment–thrust–curvature (*M–P–ϕ*) and moment–thrust–axial strain (*M–P–ε*) relations, which may be obtained separately from a moment–curvature analysis or approximated by closed-form expressions [56].
- Cross-sections are subdivided into elemental areas and the state of stresses and strains are traced explicitly using the proper stress–strain relations for all elements during the analysis.

4.7.6 Moment–Curvature Analysis

The main purpose of moment–curvature analysis is to study the section behavior. The following assumptions are usually made:

- Plane section before bending remains plane after bending.
- Shear and torsional deformations are negligible.
- Stress–strain relationships for concrete and steel are given [35].
- For reinforced concrete, a prefect bond between concrete and steel rebar exists.

The mathematical formulas used in the section analysis are (Figure 4.21)

Compatibility equations

$$\phi_x = \varepsilon/y \tag{4.33}$$

$$\phi_y = \varepsilon/x \tag{4.34}$$

Equilibrium equations

$$P = \int_A \sigma \, dA = \sum_{i=1}^{n} \sigma_i A_i \tag{4.35}$$

$$M_x = \int_A \sigma y \, dA = \sum_{i=1}^{n} \sigma_i y_i A_i \tag{4.36}$$

$$M_y = \int_A \sigma x \, dA = \sum_{i=1}^{n} \sigma_i x_i A_i \tag{4.37}$$

For a reinforced concrete member, the cross-section is divided into an appropriate number of equivalent concrete and steel elements or filaments representing the concrete and reinforcing steel as shown in Figure 4.21. Each concrete and steel layer or filament is assigned its corresponding stress–strain relationships. Confined and unconfined stress–strain relationships are used for the core concrete and for the cover concrete, respectively.

For a structural steel member, the section is divided into steel layers or filaments and a typical steel stress–strain relationship is used for tension and compact compression elements, and an equivalent stress–strain relationship with reduced yield stress and strain can be used for a noncompact compression element.

The analysis process starts by selecting a strain for the extreme concrete (or steel) fiber. Using this selected strain and assuming a section neutral axis (NA) location, a linear strain profile is constructed and the corresponding section stresses and forces are computed. Section force equilibrium is then checked for the given axial load. By changing the location of the NA, the process is repeated until

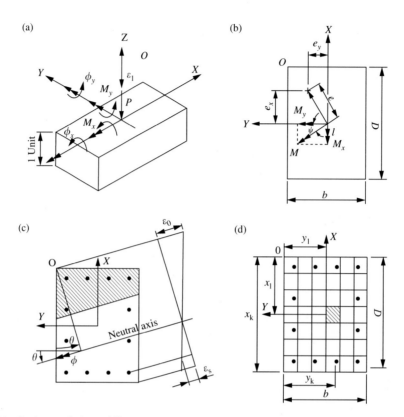

FIGURE 4.21 Section analysis modeling.

equilibrium is satisfied. Once the equilibrium is satisfied, and for the assumed strain and the given axial load, the corresponding section moment and curvature are computed by Equations 4.36 and 4.37.

A moment–curvature (M–ϕ) diagram for a given axial load is constructed by incrementing the extreme fiber strain and finding the corresponding moment and the associated curvature. An interaction diagram (M–P) relating axial load and ultimate moment is constructed by incrementing the axial load and finding the corresponding ultimate moment using the above procedure.

For a reinforced concrete section, the yield moment is usually defined as the section moment at onset of yielding of the tension reinforcing steel. The ultimate moment is defined as the moment at peak moment capacity. The ultimate curvature is usually defined as the curvature when the extreme concrete fiber strain reaches ultimate strain or when the reinforcing rebar reaches its ultimate (rupture) strain (whichever take place first). Figure 4.22a shows typical M–P–ϕ curves for a reinforced concrete section.

For a simple steel section, such as rectangular, circular-solid, and thin-walled circular section, a closed-form of M–P–ϕ can be obtained using the elastic–perfectly plastic stress–strain relations [56]. For all other commonly used steel section, numerical iteration techniques are used to obtain M–P–ϕ curves. Figure 4.22b shows typical M–P–ϕ curves for a wide-flange section.

4.7.7 Random Vibration Approach

The random vibration approach is a well-recognized and advanced seismic-response analysis method for linear multisupport-structural systems and long-span structures [57,58]. This approach provides a statistical measure of the response, which is not controlled by an arbitrary choice of the input motions, and also significantly reduces the response evaluation to that of a series of linear one-degree systems in

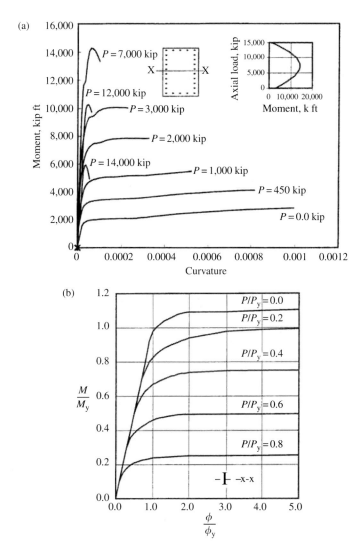

FIGURE 4.22 Typical moment–thrust–curvature curves: (a) reinforced concrete section and (b) steel section.

a way that fully accounts for the multiple-support input and the space–time correlation structure of the ground motion [58].

Although the random vibration approach has been adopted by the Eurocode [59] and widely used by Chinese engineers, it has not been accepted as a practical method of analysis for complex long-span structures by U.S. practicing engineers due to computation difficulties [57].

During 1990s, Lin and coworkers [60–64] have developed a new series of algorithms, that is, the pseudo-excitation method (PEM) series, on structural stationary/nonstationary random response analysis. The PEM is an accurate and extremely efficient method to solve complicated random vibration problems. The cross-correlation terms between all participating modes and between all excitations are all included in the responses. Most recently, the PEM has been used to analyze several long-span bridges in China. Cheng [65] analyzed the Hunan Yue-Yang Cable-stayed Bridge with a total length of 5700 m and a main span of 880 m, in which 2700 degrees of freedom, 15 multisupport ground motions, and 200 modes were used. Fan et al. [66] used PEM to analyze the Second Yangtze River Bridge at Nancha, a cable stay bridge with a total length of 1238 m and a main span of 628 m, in which 300 modes in the fast CQC (i.e., PEM) analysis and 12 multisupport ground motions were used. The computations showed that the "wave passage effect" may cause the differences of some demands up to 40%.

4.8   Seismic Detailing Requirements

4.8.1   ATC/MCEER Guidelines [37]

4.8.1.1   Minimum Seat Requirements

In SDRs 1 and 2, the minimum seat width is

$$N_{\min} = \left[0.1 + 0.0017L + 0.007H + 0.05\sqrt{H}\sqrt{1 + \left(2\frac{B}{L} \right)^2} \right] \frac{(1 + 1.25F_v S_1)}{\cos\alpha} \tag{4.38}$$

$$\frac{B}{L} \leq \frac{3}{8} \tag{4.39}$$

where L is the distance between joints (m), H is the height of the tallest pier between joints (m), and B is the width of the superstructure (m).

In SDRs 3 to 6, the minimum seat width is either Equation 4.38 or 1.5 times the displacement of the superstructure at the seat according to the following equation:

$$\Delta = R_d \Delta_e \tag{4.40}$$

$$R_d = \begin{cases} \left(1 - \dfrac{1}{R} \right) \dfrac{T^*}{T} + \dfrac{1}{R} & \text{for } T < T^* \\[2ex] 1 & \text{for } T \geq T^* \end{cases} \tag{4.41}$$

where Δ_e is the displacement demand from the seismic analysis.

4.8.1.2   *P–Δ* Requirements

In SDRs 3 to 6, the displacement of a pier or bent in the longitudinal and transverse directions determined by Equation 4.40 shall satisfy

$$\Delta \leq 0.25 C_s H = 0.25 \left(\frac{V_n}{W} \right) H \tag{4.42}$$

where V_n is the lateral nominal shear strength of the pier, W is the dead load of the pier, and H is the height of the pier from the point of fixity for foundation.

The basis of this requirement is that maximum displacement is such that the reduction in resisting force is limited to a 25% reduction from the lateral strength assuming no postyield stiffness. The inequality in Equation 4.42 is to keep the bridge pier from being significantly affected by *P–Δ* moments.

4.8.1.3   Minimum Displacement Capacity Requirements

For SDAP E, the following equation shall be satisfied:

$$\Delta_c \geq 1.5\Delta \tag{4.43}$$

where Δ_c is lateral displacement capacity of the pier or bent. It is defined as the displacement at which the first component reaches its maximum deformation. The factor of 1.5 on displacement demand recognizes the approximation of the modeling and analysis.

4.8.1.4   Structural Steel Design Requirements

4.8.1.4.1   *Limiting Width-to-Thickness Ratios*

In SDRs 2 to 6, the width-to-thickness ratio of compression elements of the columns in ductile MRF and single-column structures shall satisfy the limiting ratios specified in Table 4.10. Full penetration flange and web welds are required at beam-to-column connections.

TABLE 4.10 Recommended LRFD *Guidelines* — Limiting Width-to-Thickness Ratios [37]

Description of element	Width-to-thickness ratio, b/t	Limiting width-to-thickness ratio, λ_p	Limiting ratio, k
Flanges of I-shaped sections and channels in compression	$b_f/2t_f$	$\dfrac{135}{\sqrt{F_y}}$	0.3
Webs in combined flexural and axial compression	h_c/t_w	For $\dfrac{P_u}{\phi_b P_y} \leq 0.125$ $\dfrac{1365}{\sqrt{F_y}}\left(1 - \dfrac{1.54 P_u}{\phi_b P_y}\right)$ For $\dfrac{P_u}{\phi_b P_y} \leq 0.125$ $\dfrac{500}{\sqrt{F_y}}\left(2.33 - \dfrac{P_u}{\phi_b P_y}\right) \geq \dfrac{665}{\sqrt{F_y}}$	For $\dfrac{P_u}{\phi_b P_y} \leq 0.125$ $3.05\left(1 - \dfrac{1.54 P_u}{\phi_b P_y}\right)$ For $\dfrac{P_u}{\phi_b P_y} \leq 0.125$ $1.12\left(2.33 - \dfrac{P_u}{\phi_b P_y}\right) \geq 1.48$
Hollow circular sections (pipes)	D/t	$\dfrac{8950}{F_y}$	$\dfrac{200}{\sqrt{F_y}}$
Unstiffened rectangular tubes	b/t	$\dfrac{300}{\sqrt{F_y}}$	0.67
Longitudinally stiffened plates in compression	b/t	$\dfrac{145}{\sqrt{F_y}}$	0.32

Notes:
1. b_f and t_f are the width and thickness of an I-shaped section and h_c is the depth of that section and t_w is the thickness of its web.
2. Limits λ_p is for format $b/t \leq \lambda_p$.
3. Limits k is for format $b/t \leq k\sqrt{E/F_y}$.

TABLE 4.11 Recommended LRFD *Guidelines* — Limiting Slenderness Ratio [37]

Description of members	Limiting slenderness ratio Kl/r	Limiting length, L (m)
Unsupported distance for potential plastic hinge zone of columns		$\dfrac{17250 r_y}{F_y}$
Ductile compression bracing members	$\dfrac{2600}{\sqrt{F_y}}$	
Nominally ductile bracing members	$\dfrac{3750}{\sqrt{F_y}}$ limit is waived if members designed as tension-only bracing	

4.8.1.4.2 Limiting Slenderness Ratio

In SDRs 2 to 6, Table 4.11 summarizes the limiting slenderness ratio (KL/r) for various steel members. Recent studies found that more stringent requirement for slenderness ratios may be unnecessary, provided that connections are capable of conveying at least the member's tension capacity. The ratios shown in Table 4.11 reflect those relaxed limits.

4.8.1.4.3 Limiting Axial Load Ratio

High axial load in a column usually results in the early deterioration of strength and ductility.

The ratio of factored axial compression due to seismic load and permanent loads to yield strength $(A_g E_y)$ for columns in ductile MRFs and single-column structures shall not exceed 0.4 for SDR 2 and SDRs 3 to 6, respectively.

4.8.1.4.4 Plastic Rotation Capacities

In SDRs 3 to 6, the plastic rotational capacity shall be based on the appropriate performance level and may be determined from tests and a rational analysis. The maximum plastic rotational capacity θ_p should be conservatively limited to 0.035, 0.005, and 0.01 radians for life safety, operational performance, and in ground hinges and piles, respectively.

4.8.1.5 Concrete Design Requirements

4.8.1.5.1 Limiting Longitudinal Reinforcement Ratios

The ratio of longitudinal reinforcement to the gross cross-section shall not be less than 0.008 and not more than 0.04.

4.8.1.5.2 Shear Reinforcement

The shear strength shall be determined by either an implicit approach or an explicit approach. In the end regions, the explicit approach assumes that the shear-resisting mechanism is provided by the strut-tie model (explicit approach) such that

$$\phi V_s \geq V_u - (V_p + V_c) \tag{4.44}$$

$$V_p = \frac{\Lambda}{2} P_e \frac{D'}{L} \tag{4.45}$$

$$V_c = \begin{cases} 0.05\sqrt{f'_c}b_w d & \text{for plastic hinge zone} \\ 0.17\sqrt{f'_c}b_w d & \text{for outside plastic hinge zone} \end{cases} \tag{4.46}$$

$$V_s = \begin{cases} \dfrac{\pi}{2}\dfrac{A_{bh}}{s}f_{yh}D''\cot\theta & \text{for circular section} \\ \dfrac{A_v}{s}f_{yh}D''\cot\theta & \text{for rectangular section} \end{cases} \tag{4.47}$$

where

P_e = compressive axial force including seismic effects
D' = pitch circle diameter of the longitudinal reinforcement in a circular column, or the distance between the outermost layers of bars in a rectangular column
L = column length
Λ = fixity factor defined in Section 4.6.1.5
b_w = web width of the section
d = effective depth of the section
A_{bh} = area of one circular hoop/spiral rebar
A_{sh} = total area of transverse reinforcement in one layer in the direction of the shear force
D'' = centerline section diameter/width of the perimeter spiral/hoops
θ = principal crack angle defined in Section 4.6.1.5

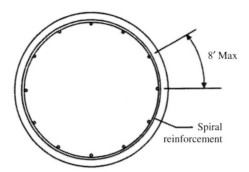

FIGURE 4.23   Column single spiral details [37].

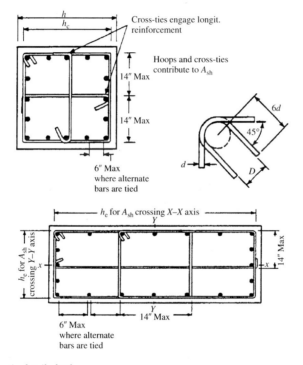

FIGURE 4.24   Column tie details [37].

4.8.1.5.3 Transverse Reinforcement in Plastic Hinge Zones

Figures 4.23 to 4.25 illustrate the typical transverse reinforcement. For confinement in plastic hinge zones, the ratio of transverse reinforcement, ρ_s, shall not be less than

$$\rho_{s\,min} = 0.008 \frac{f'_c}{U_{sf}} \left[\alpha_{shape} \left(\frac{P_e}{f'_c A_g} + \rho_t \frac{f_y}{f'_c} \right) \left(\frac{A_g}{A_{cc}} \right)^2 - 1 \right] \tag{4.48}$$

$$\rho_s = \begin{cases} \dfrac{4A_{bh}}{D's} & \text{for circular sections} \\[2ex] \dfrac{A_{sh}}{sB'} + \dfrac{A'_{sh}}{sD''} & \text{for rectangular sections} \end{cases} \tag{4.49}$$

where

A_{cc} = area of column core concrete, measured to the centerline of the perimeter hoop or spiral
A_{sh} = total area of transverse reinforcement in one layer into the direction of the applied shear

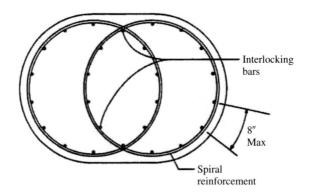

FIGURE 4.25 Column interlocking spiral details [37].

A'_{sh} = total area of transverse reinforcement in one layer perpendicular to the direction of the applied shear
B'' = core dimension of tied column in the direction of the applied shear
D'' = core dimension of tied column perpendicular to the direction of the applied shear
D' = center-to-center diameter of perimeter hoop for spiral
f_y = minimum specified yield strength of rebars
s = vertical spacing of transverse reinforcement, not exceeding 100 mm
U_{sf} = strain energy capacity of transverse reinforcement ($= 110\,\text{MPa}$)
α_{shape} = 12 and 15 for circular sections and rectangular section, respectively

To restrain buckling of longitudinal reinforcement in plastic hinges, the transverse reinforcement shall satisfy the following requirements:

For circular section

$$\rho_s = 0.016\left(\frac{D}{s}\right)\left(\frac{s}{d_b}\right)\rho_t\left(\frac{f_y}{f_{yh}}\right) \tag{4.50}$$

For rectangular section

$$A_{bh} = 0.09A_b\left(\frac{f_y}{f_{yh}}\right) \tag{4.51}$$

where D is the diameter of the circular column, d_b is the diameter of longitudinal reinforcement bars being restrained by hoop or spiral, f_y is the minimum specified yield strength of longitudinal reinforcement bars, f_{yh} is the minimum specified yield strength of transverse reinforcement bars, A_b is the area of longitudinal reinforcement bars being restrained by hoop or spiral, A_{bh} is the area of hoops or spiral or ties restraining the longitudinal steel, and s is the vertical spacing of transverse reinforcement restraining the longitudinal steel.

The transverse reinforcement shall be provided in the plastic hinge zones defined in Section 4.6.1.5. The spacing the transverse reinforcement shall meet

$$s \le \begin{cases} 6d_b \\ 0.25\,(\text{minimum member dimension}) \\ 150\,\text{mm} \end{cases} \tag{4.52}$$

4.8.1.5.4 Joint Reinforcement

- Moment-resisting integral connections shall be designed to resist the maximum plastic moment.
- The principal tension stress p_t and compression stress p_c shall be calculated by

$$\frac{p_t}{p_c} = \frac{f_h + f_v}{2} \mp \sqrt{\left(\frac{f_h + f_v}{2}\right)^2 + v_{hv}^2} \tag{4.53}$$

where f_h is the average axial stress in the horizontal direction within the plans of the connection under consideration (positive is compressive stress) (MPa), f_h is the average axial stress in the vertical direction within the plans of the connection under consideration (positive is compressive stress) (MPa), and v_{hv} is the average shear stress within the plans of the connection (MPa).

- The principal compression stress p_c shall not exceed $0.25f_c'$.
- When the principal tension stress $p_t \leq 0.29\sqrt{f_c'}$ for circular columns or columns with intersecting spirals, the volumetric ratio of transverse reinforcement in the form of spirals or hoops to be continued into the cap or footing ρ_s shall not be less than $0.29\sqrt{f_c'}/f_{yh}$ where f_{yh} is yield stress of horizontal hoop/tie reinforcement in the joint.
- When the principal tension stress $p_t > 0.29\sqrt{f_c'}$ the additional reinforcement shown in Figure 4.26 shall be provided as follows:
  - On each side column, vertical stirrups: $A_{jv} = 0.16A_{st}$.
  - Inside joint: the required vertical tie: $A_{jt} = 0.08A_{st}$.
  - Longitudinal reinforcement: $A_{jt} = 0.08A_{st}$.
  - Column hoop or spiral reinforcement into the cap: $\rho_s \geq 0.4A_{st}/l_{ac}^2$.
  where A_{st} is the total area of longitudinal steel anchored in the joint and l_{ac} is the length of column reinforcement embedded into the joint.

4.8.1.5.5 Plastic Rotation Capacities

In SDRs 3 to 6, the plastic rotational capacity shall be based on the appropriate performance level and may be determined from tests and a rational analysis. The maximum plastic rotational capacity θ_p should be conservatively limited to 0.035 (0.05), 0.01, and 0.02 radians for life safety (for liquifiable pile foundation), operational performance, and in ground hinges, respectively. For the life safety performance, the plastic hinge of a column shall be calculated by

$$\theta_p = 0.11\frac{L_p}{D'}(N_f)^{-0.5} \tag{4.54}$$

$$2 \leq N_f = 3.5(T_n)^{-1/3} \leq 10 \tag{4.55}$$

where

$N_f =$ number of cycles of loading expected at maximum displacement amplitude for liquefiable soil, take $N_f = 2$
$L_p =$ effective plastic hinge length

$$L_p = \begin{cases} 0.08\dfrac{M}{V} + 4400\varepsilon_y d_b & \text{for common columns} \\ L_g + 8800\varepsilon_y d_b & \text{for columns with isolation gap} \end{cases} \tag{4.56}$$

$\varepsilon_y =$ yield strain of longitudinal reinforcement
$L_g =$ gap length between the flare and adjacent element
$D' =$ distance between outer layers of the longitudinal reinforcement
$d_b =$ diameter of the main longitudinal reinforcement bars

4.8.2 Caltrans SDC

4.8.2.1 Minimum Seat Width Requirements

To prevent unseating of superstructures at hinges, piers, and abutments, the seat width shall be available to accommodate the anticipated thermal movement, prestressing shortening, concrete creep and

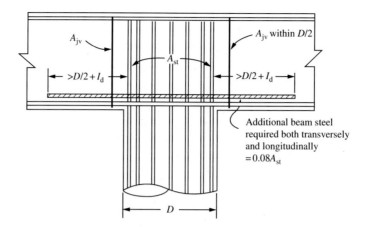

FIGURE 4.26 Additional cap beam bottom reinforcement for joint force transfer [37].
Note: I_d = development length.

shrinkage, and the relative longitudinal earthquake displacement, and shall not exceed the following minimum seat requirements [35]:

For seat width at hinges

$$N_{min} = \text{larger of} \begin{cases} \Delta_{ps} + \Delta_{cr+sh} + \Delta_{temp} + \Delta_{eq} + 100 \\ 600 \, \text{mm} \end{cases} \qquad (4.57)$$

where Δ_{eq} is total relative displacement between frames.
For seat width at abutments

$$N_{min} = \text{larger of} \begin{cases} \Delta_{ps} + \Delta_{cr+sh} + \Delta_{temp} + \Delta_{eq} + 100 \\ 760 \, \text{mm} \end{cases} \qquad (4.58)$$

where Δ_{ps}, Δ_{cr+sh}, Δ_{temp}, and Δ_{eq} are relative displacement due to prestressing, concrete creep and shrinkage, and earthquake, respectively (mm).

4.8.2.2 P–Δ Effects

The P–Δ effects tend to increase the displacement and decrease the lateral load-carrying capacity of a bridge column. These effects can typically be ignored if the moment ($P_{DL}\Delta$) is less than or equal to 20% of the column plastic moment, that is, $P_{DL}\Delta \leq 0.2M_p^{col}$.

4.8.2.3 Minimum Displacement Ductility Capacity

To ensure the dependable ductile behavior of all columns regardless of seismic demand, a minimum local displacement ductility capacity of $\mu_c = \Delta_c/\Delta_Y \geq 3$ is required and target local ductility of $\mu_c \geq 4$ is recommended. The local displacement ductility capacity shall be calculated for an equivalent member that approximates a fixed base cantilever element.

4.8.2.4 Maximum (Target) Displacement Ductility Demand

The engineers are encouraged to limit displacement ductility demands, defined as $\mu_D = \Delta_D/\Delta_Y$ to the values shown in Table 4.12.

TABLE 4.12 Caltrans Limiting Displacement Ductility Demand Values [35]

Item	Limiting $\mu_D = \Delta_D/\Delta_Y$
Single column bents supported on fixed foundations	4
Multicolumn bents supported on fixed or pinned footings	5
Pier walls (weak direction) supported on fixed or pinned footings	5
Pier walls (strong direction) supported on fixed or pinned footings	1

4.8.2.5 Minimum Lateral Strength

Although providing ductile detailing is essential for achieving the expected performance requirements, each column shall be designed to have a minimum lateral flexural to resist a lateral force of $0.1g$ or 0.1 times P_{dl}.

4.8.2.6 Structural Steel Design Requirements (Guide) [36]

4.8.2.6.1 *Limiting Width-to-Thickness Ratios*

For capacity-protected components, width–thickness ratios of compression elements shall not exceed the limiting value λ_r as specified in Table 4.13. For ductile components, width–thickness ratios shall not exceed the λ_p as specified in Table 4.13. Welds located in the expected inelastic region of ductile components are preferably complete penetration welds. Partial penetration groove welds are not recommended in these regions. If the fillet welds are only practical solution for an inelastic region, quality control and quality assurance inspection procedures for the fracture critical members shall be followed (Figure 4.27).

4.8.2.6.2 *Limiting Slenderness Ratio*

The slenderness parameter λ_c for compression members and λ_b for flexural members shall not exceed the limiting values, λ_{cp} and λ_{bp}, as specified in Table 4.14, respectively.

4.8.2.6.3 *Limiting Axial Load Ratio*

High axial load in a column usually results in the early deterioration of strength and ductility. The ratio of factored axial compression due to seismic load and permanent loads to yield strength $(A_g F_y)$ for columns in ductile moment-resisting frames and single-column structures shall not exceed 0.3.

4.8.2.6.4 *Shear Connectors*

Shear connectors shall be provided on the flanges of girders, end cross-frames, or diaphragms to transfer seismic loads from the concrete deck to the abutments or pier supports. The cross-frames or diaphragms at the end of each span are the main components to transfer the lateral seismic loads from the deck down to the bearing locations. Recent tests on a 0.4 scale experimental steel girder bridge (18.3 m long) conducted by University of Nevada, Reno [69] indicated that too few shear connectors between the girders and deck at the bridge end did not allow the end cross-frame to reach its ultimate capacity.

4.8.2.7 Concrete Design Requirements

4.8.2.7.1 *Limiting Longitudinal Reinforcement Ratios*

The ratio of longitudinal reinforcement to the gross cross-section shall not be less than 0.01 and 0.005 for columns and pier walls, respectively, and not more than 0.04.

4.8.2.7.2 *Transverse Reinforcement in Plastic Hinge Zones*

For confinement in plastic hinge zones (larger of 1.5 times cross-sectional dimension in the direction of bending and the regions of column where the moment exceeds 75% of overstrength plastic moment, M_p^{col}), transverse reinforcement shall not be less than [27]

For spiral and hoops $(\rho_s = 4A_{\text{bh}}/D's)$

$$\rho_{s,\min} = \begin{cases} 0.45\dfrac{f_c'}{f_y}\left(\dfrac{A_g}{A_c} - 1\right)\left(0.5 + \dfrac{1.25P_e}{f_c'A_g}\right) & \text{for } D \leq 0.9\,\text{m} \\[2ex] 0.12\dfrac{f_c'}{f_y}\left(0.5 + \dfrac{1.25P_e}{f_c'A_g}\right) & \text{for } D > 0.9\,\text{m} \end{cases} \tag{4.59}$$

For ties

$$A_{\text{sh,min}} = \text{larger of} \begin{cases} 0.3 sh_c\dfrac{f_c'}{f_y}\left(\dfrac{A_g}{A_c} - 1\right)\left(0.5 + \dfrac{1.25P_e}{f_c'A_g}\right) \\[2ex] 0.12 sh_c\dfrac{f_c'}{f_y}\left(0.5 + \dfrac{1.25P_e}{f_c'A_g}\right) \end{cases} \tag{4.60}$$

TABLE 4.13 Caltrans Limiting Width–Thickness Ratios [36]

No.	Description of elements	Examples	Width–thickness ratios	λ_r	λ_p
Unstiffened elements					
1	Flanges of I-shaped rolled beams and channels in flexure	Figure 4.27a Figure 4.27c	b/t	$\dfrac{370}{\sqrt{F_y}-69}$	$\dfrac{137}{\sqrt{F_y}}$
2	Outstanding legs of pairs of angles in continuous contact; flanges of channels in axial compression; angles and plates projecting from beams or compression members	Figure 4.27d	b/t	$\dfrac{250}{\sqrt{F_y}}$	$\dfrac{137}{\sqrt{F_y}}$
Stiffened elements					
3	Flanges of square and rectangular boxes and hollow structural section of uniform thickness subject to bending or compression; flange cover plates and diaphragm plates between lines of fasteners or welds	Figure 4.27b	b/t	$\dfrac{625}{\sqrt{F_y}}$	$\dfrac{290}{\sqrt{F_y}}$ (tubes) $\dfrac{400}{\sqrt{F_y}}$ (others)
4	Unsupported width of cover plates perforated with a succession of access holes	Figure 4.27d	b/t	$\dfrac{830}{\sqrt{F_y}}$	$\dfrac{400}{\sqrt{F_y}}$
5	All other uniformly compressed stiffened elements, i.e., supported along two edges	Figures 4.27a, c, d, f	b/t h/t_w	$\dfrac{665}{\sqrt{F_y}}$	$\dfrac{290}{\sqrt{F_y}}$ (w/lacing) $\dfrac{400}{\sqrt{F_y}}$ (others)
6	Webs in flexural compression	Figures 4.27a, c, d, f	h/t_w	$\dfrac{2550}{\sqrt{F_y}}$	$\dfrac{1365}{\sqrt{F_y}}$
7	Webs in combined flexural and axial compression	Figures 4.27a, c, d, f	h/t_w	$\dfrac{2550}{\sqrt{F_y}}\times\left(1-\dfrac{0.74P}{\phi_b P_y}\right)$	For $P_u \le 0.125\phi_b P_y$ $\times\dfrac{1365}{\sqrt{F_y}}\left(1-\dfrac{1.54P}{\phi_b P_y}\right)$ For $P_u > 0.125\phi_b P_y$ $\times\dfrac{500}{\sqrt{F_y}}\left(2.33-\dfrac{P}{\phi_b P_y}\right)$ $\ge\dfrac{665}{\sqrt{F_y}}$
8	Longitudinally stiffened plates in compression	Figure 4.27e	b/t	$\dfrac{297\sqrt{k}}{\sqrt{F_y}}$	$\dfrac{197\sqrt{k}}{\sqrt{F_y}}$
9	Round HSS in axial compression or flexure		D/t	$\dfrac{7930}{F_y}$	$\dfrac{8950}{F_y}$

n = number of equally spaced longitudinal compression flange stiffeners.

I_s = moment of inertia of a longitudinal stiffener about an axis parallel to the bottom flange and taken at the base of the stiffener.

Notes:

1. Width–thickness ratios shown in **Bold** are from AISC-LRFD (1999) [67] and AISC-Seismic Provisions (1997) [68]. F_y is MPa.
2. k = buckling coefficient specified by Article 6.11.2.1.3a of AASHTO-LRFD [38]. For $n=1$, $\boldsymbol{k}=(8\boldsymbol{I}_s/\boldsymbol{bt}^3)^{1/3}\le 4.0$; for $n=2, 3, 4$, and 5, $\boldsymbol{k}=(14.3\boldsymbol{I}_s/\boldsymbol{bt}^3\boldsymbol{n}^4)^{1/3}\le 4.0$.

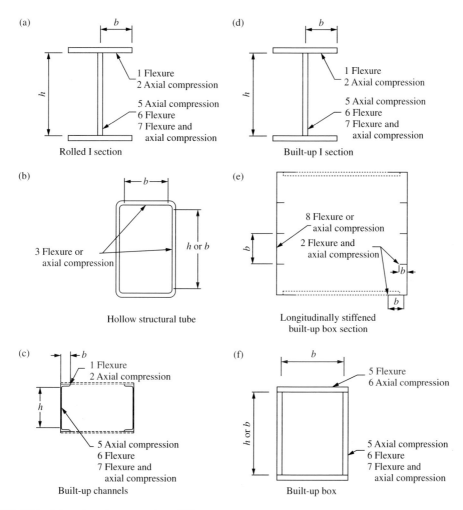

FIGURE 4.27 Selected steel cross-sections [36].

TABLE 4.14 Caltrans Limiting Slenderness Parameters [36]

Member classification		Limiting slenderness parameters
Ductile		
Compression member	λ_{cp}	0.75
Flexural member	λ_{bp}	$17240/F_y$
Capacity protected		
Compression member	λ_{cp}	1.5
Flexural member	λ_{bp}	$1970/\sqrt{F_y}$

Notes:
$\lambda_c = (KL/r\pi)\sqrt{(F_y/E)}$ (slenderness parameter for compression members).
$\lambda_b = L/r_y$ (slenderness parameter for flexural members).
λ_{cp} = limiting slenderness parameter for compression members.
λ_{bp} = limiting slenderness parameter for flexural members.
K = effective length factor of a member.
L = unsupported length of a member (mm).
r = radius of gyration (mm).
r_y = radius of gyration about the minor axis (mm).
F_y = specified minimum yield strength of steel (MPa).
E = modulus of elasticity of steel (200,000 MPa).

where A_{sh} is the total cross-sectional area of tie reinforcement including supplementary cross-tie within a section having limits of s and h_c, h_c is the core dimension of tied column in direction under consideration (out-to-out of ties), s is the vertical spacing of transverse reinforcement, and D' is the center-to-center diameter of perimeter hoop for spiral.

The spacing the transverse reinforcement in the plastic hinge zones shall meet

$$s \leq \begin{cases} 6d_b \\ 0.2\,D_{\min} \\ 220\,\text{mm} \end{cases} \tag{4.61}$$

where $D_{\min}$ is the least cross-section dimension for columns and one-half of the least cross-section dimension of piers and d_b is the diameter of the main column reinforcement.

4.8.2.7.3 Joint Proportion and Reinforcement [35,70]

- Moment-resisting integral connections shall be designed to resist the overstrength capacity $1.2 \times M_p^{\text{col}}$ and associated shear.
- The principal tension stress p_t and compression stress p_c shall not exceed $1.0\sqrt{f_c'}$ and $0.25, f_c'$, respectively. The bent cap width shall not be less than the cross-section dimension of the column in the direction of bending plus 600 mm.
- When the principal tension stress $p_t \leq 0.29\sqrt{f_c'}$ for circular columns or columns with intersecting spirals, the volumetric ratio of transverse reinforcement in the form of spirals or hoops to be continued into the cap or footing ρ_s shall not be less than $0.29\sqrt{f_c'}/f_{yh}$.

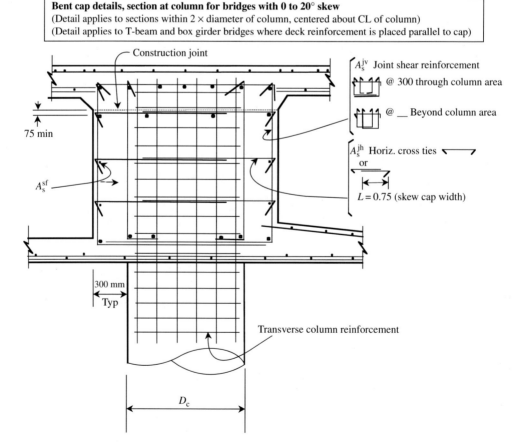

FIGURE 4.28   Example cap joint shear reinforcement — skews 0 to 20° [35].

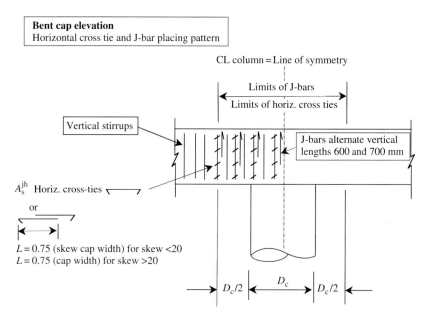

FIGURE 4.29 Location of horizontal joint shear reinforcement [35].

- When the principal tension stress $p_t > 0.29\sqrt{f_c'}$ the additional reinforcement as shown in Figure 4.28 and Figure 4.29 shall be provided and well distributed within $D/2$ from the face of the column. All joint shear reinforcement shall be as follows:
- On each side column, vertical stirrups: $A_s^{jv} = 0.2A_{st}$.
- Horizontal stirrups or tiles: $A_s^{jh} = 0.1A_{st}$.
- Longitudinal side reinforcement: $A_s^{sf} \geq 0.1A_{cap}$, where A_{cap} is area of bent cap top or bottom flexural steel.
- Column hoop or spiral reinforcement into the cap: $\rho_s \geq 0.4A_{st}/l_{ac}^2$.
- Horizontal reinforcement shall be stitched across the cap in two or more intermediate layers. The reinforcement shall be shaped as hairpins, spaced vertically at not more than 460 mm. The hairpins shall be 10% of column reinforcement. Spacing shall be denser outside the column than that used within the column.
- For bent caps skewed greater than 20°, the vertical J-bars hooked around longitudinal deck and bent cap steel shall be 8% of column steel (see Figure 4.30). The J-bars shall be alternatively 600 and 750 mm long and placed within a width of column dimension either side of the column centerline.
- All vertical column bars shall be extended as high as practically possible without interfering with the main cap bars.

4.8.2.7.4 Effective Plastic Hinge Length
The effective plastic hinge length is used to evaluate the plastic rotation of columns.

$$L_p = \begin{cases} 0.08\,L + 0.022\,f_{ye}d_b \geq 0.044\,f_{ye}d_b & \text{for columns and Type II shafts} \\ G + 0.044\,f_{ye}d_b & \text{for horizontally isolated flared columns} \\ D^* + 0.06\,H' & \text{for noncased Type I shafts} \end{cases} \qquad (4.62)$$

where f_{ye} is the expected yield stress of longitudinal reinforcement, L is the member length from the maximum moment to the point of contraflexure, H' is the length of pile shaft/column from the ground surface to point of contraflexure, above ground, and D^* is the lesser cross-section dimension of shafts.

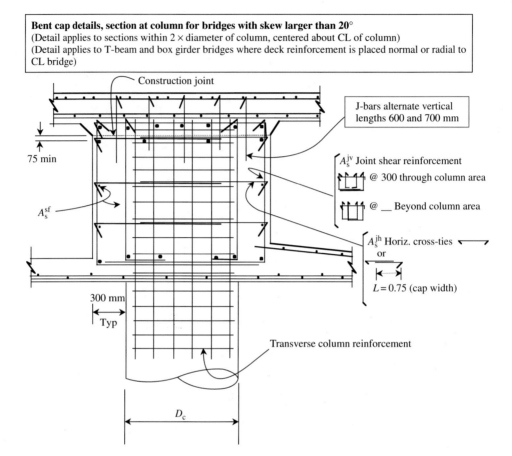

FIGURE 4.30   Example cap joint shear reinforcement — Skews > 20° [35].

4.9 Seismic Design Practice in Japan

4.9.1 Introduction

Seismic design methods for highway bridges in Japan have been developed and improved based on the lessons learned from the various past bitter experiences after the Kanto Earthquake (M7.9) in 1923. By introducing various provisions such as soil liquefaction considerations and unseating prevention devices to prevent bridges from serious damage, only a few highway bridges suffered complete collapse of superstructures in the recent past earthquakes.

However, the Hyogo-ken-Nanbu (Kobe) Earthquake (M7.3) of January 17, 1995, caused destructive damage to highway bridges. Collapse and near-collapse of superstructures occurred at nine sites and other destructive damages occurred at 16 sites [71,72]. The earthquake revealed that there were a number of critical design issues that should be reevaluated and revised in the seismic design and seismic strengthening of bridges.

Just after the earthquake, the "Committee for Investigation on the Damage of Highway Bridges Caused by the Hyogo-ken-Nanbu Earthquake" was established in the Ministry of Construction (currently Ministry of Land, Infrastructure and Transport (MLIT)) to investigate the damage and to clarify the factors that affected the damage. The committee published the intermediate investigation report in March 1995, and the final one in December 1995 [71]. Besides the investigation of damage of highway bridges, the Committee approved the "Guide Specifications for Reconstruction and Repair of Highway Bridges which Suffered Damage due to the Hyogo-ken-Nanbe Earthquake" on February 27,

1995, and the Ministry of Construction adopted on the same day that the reconstruction and repair of highway bridges that suffered damage during the Hyogo-ken-Nanbu Earthquake should be made according to the "Guide Specifications." Then, the Ministry of Construction decided on May 25, 1995, that the "Guide Specifications" should be tentatively used in all sections of Japan as emergency measures for seismic design of new highway bridges and seismic strengthening of existing highway bridges until the Design Specifications of Highway Bridges were revised.

Based on the lessons learned from the Hyogo-ken-Nanbu Earthquake through the various investigations, the seismic design specifications for highway bridges were significantly revised in 1996 [73–75]. The intensive earthquake motion with a short distance from the inland earthquakes with Magnitude 7 class as the Hyogo-ken-Nanbu Earthquake has been considered in the design.

After that, the revision works of the design specifications of highway bridges have been continuously made. The revised specifications were targeted to use the performance-based design concept and to enhance the durability of bridge structures for a long-term use, as well as the inclusion of the improved knowledge on the bridge design and construction methods after the 1996 specifications. The 2002 design specifications of highway bridges were issued by the Ministry of Land, Infrastructure and Transport on December 27, 2001, and were published with commentary from the Japan Road Association (JRA) in March 2003 [76,77].

4.9.2 Performance-Based Design Specifications

The 2002 JRA design specifications [76] are based on the performance concept for the purpose to respond the international harmonization of design codes and the flexible applications of new structures and new construction methods. The performance-based design concept is that the necessary performance requirements and the verification policies are clearly specified. The JRA specifications employ the style to specify both the requirements and the acceptable solutions including the detailed performance verification methods which are design methods and details in the 1996 design specifications [73]. For example, the analysis method to evaluate the response against the loads is placed as one of the verification methods or acceptable solutions. Therefore, designers can propose new ideas or select other design methods with the necessary verification.

The code structure of the JRA Seismic Design Specifications is as shown in Figure 4.31. The static and dynamic verification methods of the seismic performance as well as the evaluation methods of the

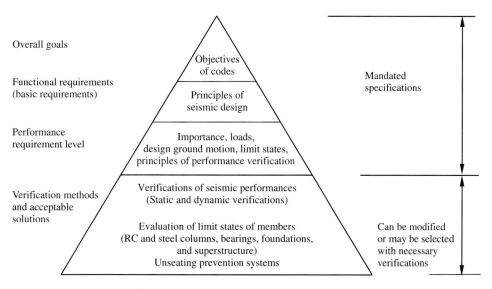

FIGURE 4.31 Code structure of JRA seismic design specification.

strength and ductility capacity of the bridge members are placed as the verification methods and the acceptable solutions, which can be modified by the designers with the necessary verifications.

4.9.3 Basic Principles of Seismic Design

Table 4.15 shows the performance matrix including the design earthquake ground motion and the seismic performance level (SPL) provided in the JRA Seismic Design Specifications. The two-level ground motions are the moderate ground motions induced in the earthquakes with high probability to occur (Level 1 earthquake) and the intensive ground motions induced in the earthquakes with low probability to occur (Level 2 earthquake).

The Level 1 earthquake provides the moderate earthquake ground motions considered in the conventional elastic design method. For the Level 2 earthquake, two types of intensive ground motions are considered. The first is the ground motions that are induced in the interplate-type earthquakes with a magnitude of around 8. The ground motion at Tokyo in the 1923 Kanto Earthquake is a typical target of this type of ground motion. The second is the ground motion developed in earthquakes with magnitude of around 7 at a very short distance. The ground motion at Kobe during the Hyogo-ken-Nanbu Earthquake is a typical target of this type of ground motion. The first and the second ground motions are named as Type-I and Type-II ground motions, respectively.

Figure 4.32 shows the acceleration response spectra of the design ground motions. It is specified that the site-specific design ground motions shall be considered if the ground motion can be appropriately estimated based on the information on the earthquake including past history and the location and detailed condition of the active faults, ground conditions including the condition from the faults to the construction sites. The site-specific design ground motion shall be developed by utilizing necessary and accurate information on the earthquake and ground conditions as well as the verified evaluation methodology of the fault-induced ground motions. However, such detailed information in the regions of Japan is very limited so far. Therefore, continuous investigation and research on this issue as well as the reflection on the practical design of highway bridges is expected.

4.9.4 Ground Motion and Seismic Performance Level

The seismic design of bridges is according to the performance matrix as shown in Table 4.15. The bridges are categorized into two groups depending on their importance: standard bridges (Type-A bridges) and important bridges (Type-B bridges). The SPL depends on the importance of bridges. For the moderate ground motions induced in the earthquakes with high probability of occurrence, both A and B bridges shall behave in an elastic manner essentially without structural damage (SPL: 1). For the intensive ground motions induced in the earthquakes with low probability of occurrence, the Type-A bridges shall prevent critical failure (SPL: 3), while the Type-B bridges shall perform with limited damage (SPL: 2).

SPLs 1 to 3 are based on the viewpoints of "safety," "functionability," and "repairability" of bridge structures during and after the earthquakes. Table 4.16 shows the basic concept of these three viewpoints of the SPL.

TABLE 4.15 Seismic Performance Matrix (Design Ground Motion and Seismic Performance Level)

Type of design ground motions	Standard bridges (Type-A)	Important bridges (Type-B)
Level 1 earthquake: ground motions with high probability of occurrence	SPL 1: prevent damage	
Level 2 Earthquake: ground motions with low probability of occurrence Interplate earthquakes (Type-I) Inland earthquakes (Type-II)	SPL 3: prevent critical damage	SPL 2: limited damage for function recovery

Note: SPL, seismic performance level.

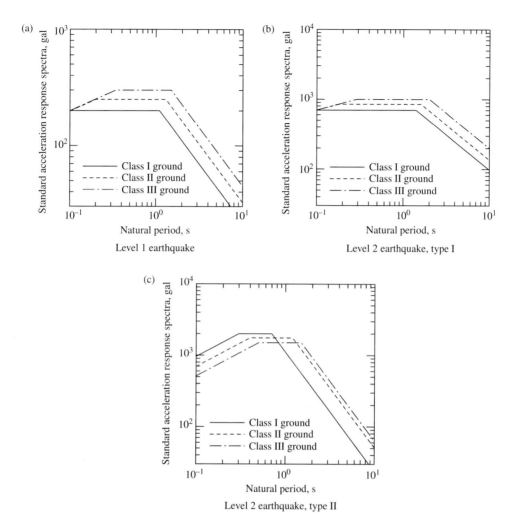

FIGURE 4.32 Design acceleration spectrum. (Class I ground: stiff ground, Class II ground: medium ground, Class III ground: soft ground.)

TABLE 4.16 Key Issues of Seismic Performance

SPL	Safety	Functionability	Repairability	
			Short term	Long term
SPL 1, Prevent damage	Safety against unseating of superstructure	Same function as before earthquake	No need of repair for function recovery	Simple repair
SPL 2, Limited damage for function recovery	Safety against unseating of superstructure	Early function recovery can be made	Function recovery can be made by temporary repair	Relatively easy permanent repair work can be made
SPL 3, Prevent critical damage	Safety against unseating of superstructure	—	—	—

4.9.5 Verification Methods of Seismic Performance

4.9.5.1 Seismic Performance Level and Limit States

In the JRA specifications [76], the determination principles of the limit state to attain the necessary seismic performance are clearly specified. For example, the basic principles to determine the limit state for SPL 2 is: (1) the plastic hinges shall be developed at the expected locations and the capacity of plastic hinges shall be determined so that the damaged members can be repaired relatively easily and quickly without replacement of main members; (2) the plastic hinges shall be developed at the portions with appropriate energy absorption and with quick repairability; (3) considering the structural conditions, the members with plastic hinges shall be appropriately arranged and the limit states of members with plastic hinges shall be determined appropriately. Based on the basic concept, the arrangement of members with plastic hinges and the limit states of members for ordinary bridge structures are specified in the commentary.

4.9.5.2 Verification Methods of Seismic Performance

It is the fundamental policy that the response of the bridge structures against design earthquake ground motions shall be verified by specified method not to exceed the specified limit states.

Table 4.17 shows the applicable verification methods of seismic performance. In the seismic design of highway bridges, it is important to increase the strength and the ductility capacity to appropriately resist the intensive earthquakes. The verification methods are based on static and dynamic analyses. In the JRA specifications, the static lateral force coefficient method with elastic design, ductility design method, and dynamic analysis are specified and these design methods shall be selected based on the configurations of bridge structures.

The static verification methods including the lateral force design method and the ductility design method are applied for the bridges with simple configuration with predominant single mode during the earthquakes. The dynamic verification method is applied for the bridges with complicated configuration; in such cases the applicability of the static verification methods is restricted. In the JRA specifications, the applicability of the dynamic analysis is expanded and the dynamic verification method is limited mainly with appropriate design consideration.

TABLE 4.17 Applicable Verification Methods of Seismic Performance Depending on Earthquake Response Characteristics of Bridge Structures

Dynamic characteristics; SPL to be verified	Bridges with simple configuration	Bridges with multi-plastic hinges and without verification of applicability of energy constant rule	Bridges with limited application of static analysis	
			With multimode response	Bridges with complicated configuration
SPL 1	Static verification	Static verification	Dynamic verification	Dynamic verification
SPL 2/SPL 3	Static verification	Dynamic verification	Dynamic verification	Dynamic verification
Example of bridges	Other bridges	(1) Bridges with rubber bearings to distribute inertia force of superstructures (2) Seismically isolated bridges (3) Rigid frame bridges (4) Bridges with steel columns	(1) Bridges with long natural period (2) Bridge with high piers	(1) Cable-stayed bridges, suspension bridges (2) Arch bridges (3) Curved bridges

Glossary

Capacity-protected component — A component expected to experience minimum damage and to behave essentially elastic during the design earthquakes.

Concentrically braced frame (CBF) — A diagonally braced frame in which all members of the bracing system are subjected primarily to axial forces.

Connections — A combination of joints used to transmit forces between two or more members.

Design earthquake — Earthquake loads represented by acceleration response spectrum (ARS) curves specified in design specifications or codes.

Displacement ductility — Ratio of ultimate-to-yield displacement.

Ductile component — A component expected to experience repairable damage during the FEE and significant damage but without failure during the SEE.

Ductility — Ratio of ultimate-to-yield deformation.

Eccentrically braced frame (EBF) — A diagonally braced frame that has at least one end of each bracing member connected to a link.

Expected nominal strength — Nominal strength of a component based on its expected yield strength.

Functional evaluation earthquake (FEE) — A lower level design earthquake that has relatively small magnitude but may occur several times during the life of the bridge. It may be assessed either deterministically or probabilistically. The determination of this event is to be reviewed by a Caltrans-approved consensus group.

Joint — An area where member ends, surfaces, or edges are attached.

Link — In EBF, the segment of a beam that is located between the ends of two diagonal braces or between the end of a diagonal brace and a column. Under lateral loading, the link deforms plastically in shear thereby absorbing energy. The length of the link is defined as the clear distance between the ends of two diagonal braces or between the diagonal brace and the column face.

Liquefaction — Seismically induced loss of shear strength in loose, cohesionless soil that results from a build up of pour pressure as the soil tries to consolidate when exposed to seismic vibrations.

Maximum credible earthquake (MCE) — The largest earthquake that is capable of occurring along an earthquake fault, based on current geologic information as defined by the 1996 Caltrans Seismic Hazard Map.

Moment-resisting frame (MRF) — A frame system in which seismic forces are resisted by shear and flexure in members and connections in the frame.

Nominal strength — The capacity of a component to resist the effects of loads, as determined by computations using specified material strength, dimensions, and formulas derived form acceptable principles of structural mechanics or by field tests or laboratory test of scaled models, allowing for modeling effects, and differences between laboratory and field conditions.

Overstrength capacity — The maximum possible strength capacity of a ductile component considering actual strength variation between the component and adjacent components. It is estimated by an overstrength factor of 1.2 times expected nominal strength.

Plastic hinge — A concentrated "zero length" hinge that maintains its plastic moment capacity and rotates freely.

Plastic hinge zone — A region of structural components that are subject to potential plastification and thus must be detailed accordingly.

Seismic performance criteria — The levels of performance in terms of postearthquake service and damage that are expected to result from specified earthquake loadings.

Safety evaluation earthquake (SEE) — An upper level design earthquake that has only a small probability of occurring during the life of the bridge. It may be assessed either deterministically or probabilistically. The deterministic assessment corresponds to the maximum credible earthquake. The probabilistically assessed earthquake typically has a long return period (approximately 1000 to 2000 years).

Ultimate displacement — The lateral displacement of a component or a frame corresponding to the expected damage level, not to exceed the displacement when the lateral resistance degrades to a minimum of 80% of the peak resistance.

Upper bound solution — A solution calculated on the basis of an assumed mechanism that is always at best equal to or greater than the true ultimate load.

Yield displacement — The lateral displacement of a component or a frame at the onset of forming the first plastic hinge.

References

[1] Moehle, J.P. and Eberhard, M.O. Chapter 34: Earthquake Damage to Bridges, *Bridge Engineering Handbook*, ed., Chen, W.F. and Duan, L., CRC Press, Boca Raton, FL, 2000.

[2] Yashinsky, M. Chapter 29a: Earthquake Damage to Structures, *Structural Engineering Handbook CRCnetBase 2000*, ed., Chen, W.F. and Duan, L., CRC Press, Boca Raton, FL, 2000.

[3] Housner, G.W. *Competing Against Time*, Report to Governor George Deuknejian from The Governor's Broad of Inquiry on the 1989 Loma Prieta Earthquake, Sacramento, CA, 1990.

[4] Caltrans, *The First Annual Seismic Research Workshop*, Division of Structures, California Department of Transportation, Sacramento, CA, 1991.

[5] Caltrans, *The Second Annual Seismic Research Workshop*, Division of Structures, California Department of Transportation, Sacramento, CA, 1993.

[6] Caltrans, *The Third Annual Seismic Research Workshop*, Division of Structures, California Department of Transportation, Sacramento, CA, 1994.

[7] Caltrans, *The Fourth Caltrans Seismic Research Workshop*, Engineering Service Center, California Department of Transportation, Sacramento, CA, 1996.

[8] Caltrans, *The Fifth Caltrans Seismic Research Workshop*, Engineering Service Center, California Department of Transportation, Sacramento, CA, 1998.

[9] FHWA and Caltrans, *The Proceedings of First National Seismic Conference on Bridges and Highways*, San Diego, CA, 1995.

[10] FHWA and Caltrans, *The Proceedings of Second National Seismic Conference on Bridges and Highways*, Sacramento, CA, 1997.

[11] Kawashima, K. and Unjoh, S., The Damage of Highway Bridges in the 1995 Hyogo-Ken Naubu Earthquake and its Impact on Japanese Seismic Design, *J. Earthquake Eng.*, 1(2), 1997, 505.

[12] Park, R., Ed., Seismic Design and Retrofitting of Reinforced Concrete Bridges, *Proceedings of the Second International Workshop*, Queenstown, New Zealand, August, 1994.

[13] Astaneh-Asl, A. and Roberts, J., Ed., Seismic Design, Evaluation and Retrofit of Steel Bridges, *Proceedings of the First U.S. Seminar*, San Francisco, CA, 1993.

[14] Astaneh-Asl, A. and Roberts, J., Ed., Seismic Design, Evaluation and Retrofit of Steel Bridges, *Proceedings of the Second U.S. Seminar*, San Francisco, CA, 1997.

[15] Housner, G.W. *The Continuing Challenge — The Northridge Earthquake of January 17, 1994*, Report to Director, California Department of Transportation, Sacramento, CA, 1994.

[16] Priestley, M.J.N., Seible, F., and Calvi, G.M., *Seismic Design and Retrofit of Bridges*, John Wiley and Sons, New York, NY, 1996.

[17] Caltrans, *The Sixth Caltrans Seismic Research Workshop*, Divisions of Engineering Services, California Department of Transportation, Sacramento, CA, June 12–13, 2001.

[18] FHWA-NSF, *Proceedings of 16th US–Japan Bridge Engineering Workshop*, Lake Tahoe, Nevada, October 2–4, 2000.

[19] ATC, *Seismic Design Guidelines for Highway Bridges*, Report No. ATC-6, Applied Technology Council, Redwood City, CA, 1981.

[20] AASHTO, *Standard Specifications for Highway Bridges*, 17th ed., American Association of State Highway and Transportation Officials, Washington, DC, 2002.

[21] Caltrans, *Bridge Design Specifications*, California Department of Transportation, Sacramento, CA, 1990.

[22] Caltrans, *Bridge Design Specifications, LFD Version*, California Department of Transportation, Sacramento, CA, 2000.

[23] Caltrans, *Bridge Memo to Designers (20-1) — Seismic Design Methodology*, California Department of Transportation, Sacramento, CA, January 1999.

[24] Caltrans, *Bridge Memo to Designers (20-4)*, California Department of Transportation, Sacramento, CA, 1995.

[25] Caltrans, *Bridge Memo to Designers (20-11) — Establishing Bridge Seismic Design Criteria*, California Department of Transportation, Sacramento, CA, January 1999.

[26] Caltrans, *Bridge Design Aids*, California Department of Transportation, Sacramento, CA, 1995.

[27] Caltrans, *Bridge Design Specifications — LFD Version*, California Department of Transportation, Sacramento, CA, April 2000.

[28] Caltrans, *San Francisco — Oakland Bay Bridge West Spans Seismic Retrofit Design Criteria*, Prepared by Reno, M. and Duan, L. Edited by Duan, L., California Department of Transportation, Sacramento, CA, 1997.

[29] Caltrans, *San Francisco-Oakland Bay Bridge East Span Seismic Safety Project Design Criteria, Version 7*, Prepared by T.Y. Lin, Moffatt & Nichol Engineers, California Department of Transportation, Sacramento, CA, July 6, 1999.

[30] IAI, *Benicia-Martinez Bridge Seismic Retrofit — Main Truss Spans Final Retrofit Strategy Report*, Imbsen and Association, Inc., Sacramento, CA, 1995.

[31] FHWA, *Seismic Design and Retrofit Manual for Highway Bridges*, Report No. FHWA-IP-87-6, Federal Highway Administration, Washington, DC, 1987.

[32] FHWA, *Seismic Retrofitting Manual for Highway Bridges*, Publication No. FIIWA-RD-94-052, Federal Highway Administration, Washington, DC, 1995.

[33] ATC, *Improved Seismic Design Criteria for California Bridges: Provisional Recommendations*, Report No. ATC-32, Applied Technology Council, Redwood City, CA, 1996.

[34] Rojahn, C. et al. *Seismic Design Criteria for Bridges and Other Highway Structures*, Report NCEER-97-0002, National Center for Earthquake Engineering Research, State University of New York at Buffalo, Buffalo, NY, 1997. Also refer as ATC-18, Applied Technology Council, Redwood City, CA, 1997.

[35] Caltrans, *Seismic Design Criteria, Version 1.3*, California Department of Transportation, Sacramento, CA, February 2004.

[36] Caltrans, *Guide Specifications for Seismic Design of Steel Bridges*, 1st ed., California Department of Transportation, Sacramento, CA, December 2001.

[37] ATC/MCEER Joint Venture, *Recommended LRFD Guidelines for the Seismic Design of Highway Bridges, Part I: Specifications; Part II: Commentary and Appendixes, Preliminary Report*, ATC Report Nos. ATC-49a and ATC-49b, Applied Technology Council, Redwood City, California and MCEER Technical Report No. MCEER-02-SP01, Multidisciplinary Center for Earthquake Engineering Research, State University of New York at Buffalo, Buffalo, NY, November 2001.

[38] AASHTO, *LRFD Bridge Design Specifications*, 3rd ed., American Association of State Highway and Transportation Officials, Washington, DC, 2004.

[39] Chen, W.F. and Duan, L., Ed., *Bridge Engineering Handbook*, CRC Press, Boca Raton, FL, 2000.

[40] Priestley, N., Myths and Fallacies in Earthquake Engineering — Conflicts Between Design and Reality, *Proceedings of Tom Paulay Symposium — "Recent Development in Lateral Force Transfer in Buildings,"* University of California, San Diego, CA, 1993.

[41] Kowalsky, M.J., Priestley, M.J.N., and MacRae, G.A., *Displacement-Based Design*, Report No. SSRP-94/16, University of California, San Diego, CA, 1994.

[42] Troitsky, M.S. Chapter 1: Conceptual Bridge Design, *Bridge Engineering Handbook*, ed., Chen, W.F. and Duan, L., CRC Press, Boca Raton, FL, 2000.

[43] Keever, M.D., Caltrans Seismic Design Criteria, *Proceedings of 16th U.S.–Japan Bridge Engineering Workshop*, Lake Tahoe, Nevada, October 2–4, 2000.

[44] Cheng, C.T. and Mander, J.B. *Seismic Design of Bridge Columns Based on Control and Repairability of Damage*, NCEER-97-0013, State University of New York at Buffalo, Buffalo, NY, 1997.

[45] USGS, http://www.geohazards.cr.usgs.gov/eq/, 2001.

[46] Mander, J.B. and Cheng, C.T. *Seismic Resistance of Bridge Piers Based on Damage Avoidance Design*, NCEER-97-0014, State University of New York at Buffalo, Buffalo, NY, 1997.

[47] Duan, L. and Cooper, T.R. Displacement Ductility Capacity of Reinforced Concrete Columns, *ACI Concrete Int.*, 17(11), 1995, 61–65.

[48] Park, R., and Paulay, T. *Reinforced Concrete Structures*, John Wiley and Sons, New York, NY, 1975.

[49] Akkari, M. and Duan, L., Chapter 36: Nonlinear Analysis of Bridge Structures, *Bridge Engineering Handbook*, ed., Chen, W.F. and Duan, L., CRC Press, Boca Raton, FL, 2000.

[50] Caltrans, *Seismic Design of Abutments for Ordinary Standard Bridges*, Division of Structure Design, California Department of Transportation, Sacramento, CA, March 20, 2001.

[51] AASHTO, *Guide Specifications for Seismic Isolation Design*, American Association of State Highway and Transportation Officials, Washington, DC, 2000.

[52] Zhang, R., Chapter 41: Seismic Isolation and Supplemental Energy Dissipation, *Bridge Engineering Handbook*, ed., Chen, W.F. and Duan, L., CRC Press, Boca Raton, FL, 2000.

[53] King, W.S., White, D.W., and Chen, W.F. Second-Order Inelastic Analysis Methods for Steel-Frame Design, *J. Struct. Eng.*, ASCE, 118(2), 1992, 408–428.

[54] Levy, R., Joseph, F., and Spillers, W.R. Member Stiffness with Offset Hinges, *J. Struct. Eng.*, ASCE, 123(4), 1997, 527–529.

[55] Chen, W.F. and Toma, S. *Advanced Analysis of Steel Frames*, CRC Press, Boca Raton, FL, 1994.

[56] Chen, W.F. and Atsuta, T. *Theory of Beam-Columns*, Vols 1 and 2, McGraw-Hill Inc., New York, NY, 1977.

[57] Kiureghian, A.D. and Neuenhofer, A. Response Spectrum Method for Multi-Support Seismic Excitations. *Earthquake Eng. Struct. Dyn.*, 21, 1992, 713–740.

[58] Heredia-Zavoni, E. and Vanmarcke, E.H. Seismic Random-Vibration Analysis of Multisupport-Structural Systems, *J. Eng. Mech.*, ASCE, 120(5), 1994, 1107–1128.

[59] European Committee for Standardization, *Eurocode 8: Structures in Seismic Regions — Design Part 2: Bridges*, 1995.

[60] Lin, J.H., Zhang, W.S., and Williams, F.W. Pseudo-Excitation Algorithm for Nonstationary Random Seismic Responses, *Eng. Struct.*, 16, 1994, 270–276.

[61] Lin, J.H., Zhang, W.S., and Li, J.J. Structural Responses to Arbitrarily Coherent Stationary Random Excitations, *Comput. Struct.*, 50, 1994, 629–633.

[62] Lin, J.H., Li, J.J., Zhang, W.S., and Williams, F.W. Non-stationary Random Seismic Responses of Multi-support Structures in Evolutionary Inhomogeneous Random Fields, *Earthquake Eng. Struct. Dyn.*, 26, 1997, 135–145.

[63] Lin, J.H., Zhao, Y., and Zhang, Y.H., Accurate and Highly Efficient Algorithms for Structural Stationary/Non-stationary Random Responses, *Comput. Meth. Appl. Mech. Eng.*, 191(1–2), 2001, 103–111.

[64] Lin, J.H., Zhang, Y.H., and Zhao, Y., High Efficiency Algorithm Series of Random Vibration and Their Applications, *Proceedings of the 8th International Conference on Enhancement and Promotion of Computational Methods in Engineering and Science*, Shanghai, China, 2001, pp. 120–131.

[65] Cheng, W., Spectrum Simulation Models for Random Ground Motions and Analysis of Long-Span Bridges under Random Earthquake Excitations, *Doctoral Dissertation*, Hunan University, Changsha, China, 2000.

[66] Fan, L.C., Wang, J.J., and Chen, W., Response Characteristics of Long-Span Cable-Stayed Bridges under Non-uniform Seismic Action, Dept of Bridge, *Chinese J. Comput. Mech.*, 18(3), 2001, 358–363.

[67] AISC, *Load and Resistance Factor Design Specification for Structural Steel Buildings*, 3rd ed., American Institute of Steel Construction, Chicago, IL, 1999.

[68] AISC, *Seismic Provisions for Structural Steel Buildings (1997), Supplement No. 1* (1999) and *Supplement No. 2* (2000), American Institute of Steel Construction, Chicago, IL, 1997.

[69] Carden, L., Garcia-Alvarez, S., Itani, A., and Buckle, I., Cyclic Response of Steel Plate Girder Bridges in the Transverse Direction, *The Sixth Caltrans Seismic Research Workshop*, Divisions of Engineering Services, California Department of Transportation, Sacramento, CA, June 12–13, 2001.

[70] Zelinski, R. *Seismic Design Momo Various Topics Preliminary Guidelines*, California Department of Transportation, Sacramento, CA, 1994.

[71] Ministry of Construction. *Report on the Damage of Highway Bridges by the Hyogo-ken Nanbu Earthquake*, Committee for Investigation on the Damage of Highway Bridges Caused by the Hyogo-ken Nanbu Earthquake, 1995 (in English).

[72] Kawashima, K. *Impact of Hanshin/Awaji Earthquake on Seismic Design and Seismic Strengthening of Highway Bridges*, Report No. TIT/EERG 95-2, Tokyo Institute of Technology, 1995 (in English).

[73] Japan Road Association, *Design Specifications of Highway Bridges, Part I Common Part, Part II Steel Bridges, Part III Concrete Bridges, Part IV Foundations, and Part V Seismic Design*, 1996 (in Japanese, Part. V: English version, July 1998).

[74] Kawashima, K. et al. *1996 Design Specifications for Highway Bridges*, 29th UJNR Joint Panel Meeting, May 1996 (in English).

[75] Unjoh, S. Chapter 44: Seismic Design Practice in Japan, *Bridge Engineering Handbook*, ed., Chen, W.F. and Duan, L., CRC Press, Boca Raton, FL, 2000.

[76] Japan Road Association, *Design Specifications of Highway Bridges, Part I Common Part, Part II Steel Bridges, Part III Concrete Bridges, Part IV Foundations, and Part V Seismic Design*, 2002 (in Japanese, Part. V: English version, June 2003).

[77] Unjoh, S. et al. *Design Specifications for Highway Bridges*, 34th UJNR Joint Panel Meeting, May 2002.

5

Performance-Based Seismic Design and Evaluation of Building Structures

Sashi K. Kunnath
*Department of Civil and
Environmental Engineering,
University of California,
Davis, CA*

Performance-based design (PBD) has emerged as the new paradigm in seismic engineering. The concept of PBD is not limited to the field of seismic design, though the material covered in this chapter focuses primarily on recent developments in earthquake engineering. Performance-based seismic engineering (PBSE) is still an evolving methodology; hence, the information presented in this chapter should not be interpreted as an existing design specification or an adopted standard. Rather, the contents of this chapter should be viewed as an introduction to an emerging concept for seismic design of building systems.

There are several completed and ongoing efforts to develop performance-based seismic design methodologies. Published documents such as ATC-40 (1996), FEMA-350 (2000), and FEMA-356 (2000) embody key aspects of PBD; however, they were each developed with limited objectives. FEMA-350 applies to new steel moment frames only. ATC-40 and FEMA-356 contain guidelines for the evaluation and rehabilitation of existing buildings; the former is limited to the evaluation of existing reinforced concrete buildings, while the latter covers all building types and is, therefore, referred to as a "prestandard" suggesting that it may resemble the model of a PBD code.

The purpose of this chapter is to provide readers with an overview of existing procedures that attempt to incorporate performance-based concepts in seismic design of new structures or evaluation of existing structures. Several potential approaches to accomplishing the goals of PBD will be discussed and an illustrative example that applies each conceptual framework in the evaluation of a hypothetical design will be presented so as to offer a basis for comparing the relative merits and potential drawbacks of each approach. Much of the emphasis will be placed on buildings since they constitute the largest stock of structures in the built environment, though the procedures outlined here can be extended with minimal effort to other structural systems.

5.1   Some Issues in Current Seismic Design

The first question that comes to mind when introducing a new methodology is the obvious one: why do we need a new procedure? What is inherently wrong or inadequate in the existing provisions for design that warrants a new look at the entire process? Hence, the task of introducing PBSE is more easily accomplished by highlighting the limitations and drawbacks of existing seismic design procedures.

Since the early development of seismic design codes, global response modification factors (or R-factors) have remained at the core of seismic force formulas. The main purpose of the force-reduction factors used in seismic design is to simplify the analysis process so that elastic methods can be used to approximately predict the expected inelastic demands in a structure subjected to the design loads. They account for reductions in seismic force values due to a variety of factors including system inherent ductility, overstrength, and redundancy. Of these, only the ductility component of the R-factor is generally implied in the design provisions because systems with larger expected ductility have the lowest reduction factors. Current codes also specify a displacement amplification factor C_d that quantifies the expected inelastic displacement of the system. Both R and C_d factors are global response measures that do not provide an assessment of structural performance at the component level. There is growing awareness that force-based design using R and C_d factors has serious shortcomings. For instance, these factors are independent of the building period and ground motion characteristics. Additionally, the same R-factor is used for moment-resisting reinforced concrete (RC), steel, and braced frames. It is clear that a single global response modifier cannot capture the progressive distribution of nonlinearities between various structural elements, the resulting redistribution of seismic demands inside the structure, and the changes that occur during the course of the seismic motion. In addition to differences in seismic demands and failure mechanisms, the damage distribution is also likely to vary from one structure to the other even though they have all been designed for the same R-value. A coherent description of the meaning and basis of establishing force-reduction factors is outlined in two papers by Uang (1991) and Uang and Bertero (1991). Figure 5.1 shows the base-shear versus roof displacement response of a typical building structure. The vertical axis in Figure 5.1 shows the base shear coefficient, which is the total shear normalized by the seismic weight of the building (V/W). The design base shear coefficient of the building is C_s, while the corresponding elastic strength is C_e. As is evident from the response of the structure, the yield strength of the building is C_y (assuming a bilinear idealization as shown). First yielding in a member in the system should typically commence at C_s, though material overstrength and member sizing

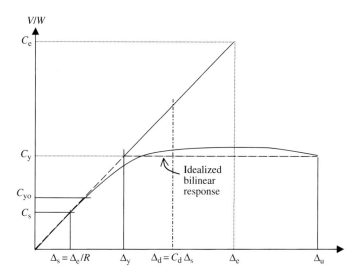

FIGURE 5.1   Conceptual basis behind use of *R*-factors.

may delay initial yielding beyond this value. Hence, the actual force-reduction factor or response modification factor as defined in IBC (2000) can be defined as follows:

$$R = \frac{C_e}{C_s} = \frac{C_e}{C_y} \cdot \frac{C_y}{C_s} \qquad (5.1)$$

The ratio C_e/C_y can be viewed as the response modification factor related to ductility (R_μ), while the ratio C_y/C_s contains two components: overstrength (R_Ω) and redundancy (R_R). Hence, the *R*-factor can be broken down into three main components as indicated in Equation 5.2:

$$R = R_\mu R_\Omega R_R \qquad (5.2)$$

Numerous factors influence each of the modifiers that appear in Equation (5.2). For example, the postyield strain (or stiffness) can affect both R_μ and R_Ω. The use of higher material strengths than those specified in the design, satisfying minimum code requirements for detailing, etc., the presence of nonstructural components, and the oversizing of members have an impact primarily on R_Ω and R_R. The product $R_\Omega R_R$ is likely to have greater variability for RC than for steel structures. It is clearly difficult to isolate the different components of the reduction factor and thereby provide engineers with an understanding of the demands imposed not only on the overall system but also on individual components in the system and the margin of safety against failure.

Since the total lateral force or base shear is the primary design parameter, the current code format is regarded as a "strength-based" design procedure. The idea of distributing strength throughout the structure rather than relying on a single base-shear parameter is recognized in the concept of capacity design (Park and Paulay 1976), which forms the basis of the New Zealand building code. The realization that displacements are more critical than forces initiated the move toward displacement-based design (Moehle 1992, 1996; Priestley and Calvi 1997; Chopra and Goel 2001). Procedures to enhance displacement-based methods yet retain the simplicity of a response spectrum to characterize the hazard were developed by Fajfar and Gasperic (1996). Other interesting research on the shortcomings of existing procedures and the promise of new ideas to advance seismic design methodologies was presented at a workshop in Slovenia (Fajfar and Krawinkler 1997).

While displacement (or deformation) based criteria still remain at the core of ongoing developments to improve the state of practice, the need to correlate measures of performance with measures of demand has ushered the need for performance-based principles to be integrated into the design process. Following the observed damage in the 1989 Loma Prieta earthquake and the aftermath of the seismic events in Northridge and Kobe, a national effort was launched to reevaluate the principles governing modern seismic design, which culminated in the release of FEMA-356. Two developments that laid the foundation for the Federal Emergency Management Agency (FEMA) document include the prior effort that led to ATC 3-06 (1978) and a parallel effort that produced Vision 2000 (SEAOC 1995). ATC-3-06 marked a significant improvement in seismic provisions with the recognition of a seismic hazard in terms of a mean annual frequency (or return period). The specified return period of 475 years (or a 10% probability of exceedance in 50 years) and associated information (effective peak accleration (EPA), effective peak velocity (EPV)) facilitated the development of a design spectrum. Vision 2000, an effort initiated and supported by the Structural Engineers Association of California (SEAOC), was concerned with developing a framework for seismic design that explicitly addressed issues beyond life safety to include damageability and functionality.

5.2   Introduction to PBSE

The concepts of hazard and performance level are implied in the provisions of standard building codes such as International Building Code (IBC). The *SEAOC Blue Book*, regarded as the commentary for the Uniform Building Code (now IBC) provisions, clearly identifies the following performance objectives:

- No damage for minor levels of ground motion.
- No structural damage but some nonstructural damage for moderate ground motions.
- Structural and nonstructural damage but no collapse for major levels of ground motion (which include the strongest credible earthquake at the site).

Despite the recognition of three damage states implied in modern codes, life safety has historically been the principle concern in seismic design. Furthermore, there are no explicit procedures or processes that allow an engineer to evaluate the expected performance of the final design or assess the margin of safety provided by satisfying code requirements. Recent experience has shown that property damage and related losses resulting from minor to moderate earthquakes is significant. Nonstructural considerations and business losses can dominate the cost–benefit ratio in a life-cycle cost analysis (Krawinkler 1997). Hence, emphasis on nonstructural issues and business interruption losses must become part of the equation in a holistic treatment of seismic design. This and other concerns discussed in the previous section have led researchers and engineers to rethink the principles and process governing modern seismic design. The various alternative strategies proposed by numerous individual and group research efforts have collectively led to the evolution of performance-based concepts that allow the design of buildings (or the rehabilitation of existing buildings) with due consideration to different design objectives and varying levels of risk and loss. While current design methodology relies considerably on empirical formulations and prescriptive procedures that require only minimal design checks (such as drift limits following the sizing of sections), PBD methodology calls for a detailed demand evaluation of the building model under simulated earthquake loads to assess if the design objective has been achieved. The assessment is intended to project the potential damage and associated losses, which in turn allows all stakeholders (from the building owner to insurers and the building occupants) to specify desired levels of performance, which then becomes the basis of design for the structural engineer. While this entails a more significant responsibility on the part of the design professional, it also means that greater flexibility is possible in the design — in terms of new design alternatives and techniques (the use of seismic protection devices is one example) that can be utilized to achieve the expected performance objective.

5.2.1   Elements of a Typical Performance-Based Methodology

The first definitive document that laid the basis for PBD in seismic engineering is FEMA-356 (2000), which essentially synthesizes two earlier reports, FEMA-273 and FEMA-274, and deals with seismic rehabilitation of existing buildings. A parallel effort that resulted in the publication of ATC-40 (1996) is limited to RC buildings but is more comprehensive in its treatment. At least one other guideline that builds on FEMA-273 is the result of the FEMA-sponsored SAC project that produced FEMA-350 (2000). Other notable research that addresses issues in performance-based engineering includes ongoing efforts at the Mid-America Earthquake (MAE) and Pacific Earthquake Engineering Research (PEER) centers.

The contents of this chapter are based on the concepts and procedures outlined in the various documents and guidelines cited above. All performance-based methodologies share common elements. Though the details of implementation may vary from one methodology to the next, the basic objectives of the each approach are essentially the same. An attempt is made in the following sections to classify the key components of a typical PBD methodology in three collective steps:

Step 1. a. Define a performance objective that incorporates a description of both the hazard and the expected level of performance.
       b. Select a trial design.
Step 2. Determine seismic demands on the system and its components through an analysis of a mathematical model of the structure.
Step 3. Evaluate performance (at the system and component levels) to verify if the performance objective defined in Step 1 has been met. If performance levels are not satisfactory, revise design and return to Step 2.

A brief overview of the three-step process is outlined in the following sections.

5.2.1.1 Quantifying Performance Objectives

Stated simply, a performance objective specifies the desired seismic performance target for the structure. A performance objective consists of two parts: a level of performance that is typically expressed in terms of a *damage state* or *decision variable* and a *hazard level* that describes the expected seismic load at the site. Some examples of performance levels include: collapse prevention, life safety (preservation of human life), and immediate occupancy of the building following the design seismic event. Alternatively, a performance level can be expressed in terms of its economic impact (dollar loss or downtime). A commonly used description of hazard is the probability of the event in a given duration, for example, 10% probability of the design seismic loads being exceeded in 50 years. The intent of this notion of a performance objective is conveyed in Figure 5.2, which quantifies the allowable damage or loss as

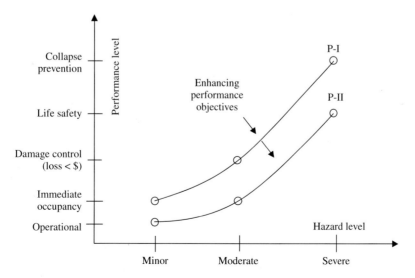

FIGURE 5.2 Defining performance objectives.

a function of the intensity of the earthquake load. As suggested earlier, a hazard level denotes the severity of the earthquake. A minor hazard level would represent an event with a higher probability of being exceeded during the life of the structure, while an increased hazard level representing a more severe event would indicate a reduced probability for the same duration. Damage states may be quantified in numerous ways, but two of the most common approaches are physical damage states based on the degree of inelastic response of members and expected losses including economic and indirect losses.

Since a structural system is composed of both primary and secondary elements, the specification of a performance level can be distinguished depending on the structural function. Hence, the structural performance level can be different from the nonstructural performance level for a given performance objective.

5.2.1.1.1 Description of Hazard

Earthquake ground motions, or the characteristics of the expected ground motions, are generally used in conjunction with a performance objective. Given the uncertainty in predicting ground motions, the task of establishing a site-specific design event is a difficult but critical part of any PBD methodology. The description of the hazard is meant to represent the seismic threat given knowledge of the potential earthquake sources at the site. This is typically accomplished through a probabilistic seismic hazard assessment that results in a hazard curve describing mean annual frequency of exceeding a certain spectral acceleration magnitude.

The specification of ground shaking is inevitably linked to two factors: (1) site geology and related site characteristics, which may be classified by soil profile or other soil parameters such as shear wave velocity, undrained shear strength, etc. and (2) site seismicity, which includes information on the seismic source. Ultimately, the design earthquake needs to be quantified in terms of a response spectrum as shown in Figure 5.3. The spectrum is characterized by three critical points: the peak ground acceleration (PGA) S_0, the region of maximum spectral demands (S_m between T_m and T_n), and finally a description of the spectral demands beyond T_n.

To allow for a range of performance objectives, it is necessary to define more than a single hazard level. The prevalent thinking in most PBD guidelines is to consider at least three levels:

- *Hazard level I (or a basic service level earthquake).* Earthquakes representing this level of hazard are expected to occur more frequently, and therefore, the likelihood of such an event during the life of the structure is high. Probabilistically speaking, this event is typically equated with a 50% probability of being exceeded in 50 years or a mean return period of 72 years.
- *Hazard level II (or a design level earthquake).* This is generally the design level event. Earthquakes at this level of hazard are normally assumed to have a 10% probability of being exceeded in 50 years.

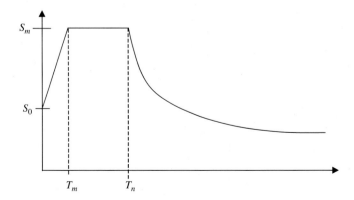

FIGURE 5.3 Key parameters in design response spectrum.

- *Hazard level III (or a maximum credible level earthquake).* Finally, it will be necessary to define a maximum credible event at the site. Such an event can be used to enable higher performance objectives (particularly for critical facilities). Here, the event is assumed to have only a 2% probability of being exceeded in 50 years.

5.2.1.2   Estimation of Demand

Once a performance objective has been defined, the next task is to determine if the design of the structure meets the stated performance objective. This means that a mathematical model of the structure is subjected to the seismic design loads (as specified by the hazard at the site in the previous step) and the demands on the system and its components are evaluated. The task of estimating seismic demand involves the determination of deformations and forces in both structural and nonstructural elements in the structure.

The prediction of deformation demands is a crucial and challenging aspect of a PBD approach. Currently, the only requirement in design that examines demand is the check on drift limitations, which is accomplished by linear elastic methods. Though some of the proposed PBD approaches suggest the use of linear methods to determine demands, there is concern that the introduction of a simplified technique into an otherwise advanced and comprehensive process will prove to be the undoing of the methodology. Errors in estimating the demand can propagate through the process and lead to misleading conclusions on the performance of the structure. Hence, for PBD methodology to become acceptable to engineers and simultaneously pass the scrutiny of researchers, there has to be consensus on an analytical approach (linear or nonlinear and static or dynamic) to evaluate seismic demands.

The process of estimating seismic demands can be subdivided into two primary tasks: developing a model of the structure wherein all structural components and essential nonstructural components are adequately represented and carrying out an analysis of the model subjected to loads that characterize the hazard level. The model may be simple or complex and is dependent on the computational tool (software) that will be used in the analysis.

The deformation demands in a structure and its elements are dependent on a variety of factors:

- *The characteristics of the seismic input.* A response spectrum is valid for static methods including modal analysis, while ground motions (recorded or simulated) are required for time history analyses.
- *Modeling of elements.*
- *Consideration of nonstructural components.*
- *Accurate estimation of material properties.*
- *Modeling of material behavior.* This relates to modeling the nonlinear response characteristics of the elements of the structure, particularly the nature of energy dissipation (hysteresis) of members. Consideration of cyclic degradation is critical to the assessment of performance and must be incorporated in the analytical model.
- *Method of analysis.* This refers to the four possible methods that can be used to estimate forces and deformations in the structural model. Static methods can be used if the seismic input is specified in terms of a response spectrum while transient methods employing a time-integration scheme must be used for ground motion input. In either case, the system of equations may be solved using a constant stiffness matrix resulting in a linear solution, or advanced techniques to incorporate material or geometric nonlinearities can be utilized.

The seismic input will already be defined in the first step in terms of either a response spectra or a set of actual ground motions that satisfy certain criteria. The modeling of structural elements that includes the specification of material-dependent force–deformation behavior is limited by the analytical tool being used to evaluate the model. Most PBD approaches utilize an element-by-element discretization of the building and specify modeling characteristics based on moment and rotation at the ends of the element with due consideration to interaction with axial and shear forces. Assuming an acceptable mathematical model of the system is developed, the next step is to carry out an analysis of the model. This brings us to a crucial phase in the process. What kind of analysis is appropriate to accurately estimate the imposed seismic demands? Since the motivation for moving away from traditional seismic design is to abandon linear force based methods, it can be argued that the analysis has to be nonlinear. In this context, pushover methods or nonlinear static approaches have gained considerable popularity since they avoid the difficulties and uncertainties associated with time-history analyses. However, it must not be overlooked that an earthquake is a dynamic event and the reliability of using static methods to predict dynamic behavior must be carefully evaluated before prescribing its use in a performance-based evaluation. FEMA-356 prescribes the use of both linear and nonlinear methods, though certain restrictions are placed on linear and static approaches. ATC-40 restricts the analytical solution to nonlinear methods only. All four methods of analysis are discussed later in this chapter, including limitations and potential issues on their application in PBSE.

5.2.1.3   Assessment of Performance

This can be regarded as the third and final phase of a performance-based evaluation. If the estimation of seismic demands is a critical phase in a PBD methodology that must be carried out cautiously with attention to detail, then the next stage in a performance-based evaluation wherein the estimated demands are transformed into performance measures is probably the most contentious. In this phase of the evaluation, demand values such as displacement, drift, and plastic rotation need to be interpreted on a damage scale D $(0 < D < 1)$ that encompasses the complete range from elastic undamaged response to near or total collapse.

The question that needs to be addressed here is, what is the most convenient yet rational manner to translate response quantities from a structural analysis into qualitative measures of performance? Much of the work that has been conducted in the area of damage modeling becomes relevant in this context. Currently, only the PEER methodology (to be discussed later in this chapter) uses damage measures explicitly in its formulation. The approach that is favored by engineers (if FEMA-356 is used as a frame of reference) is the use of so-called acceptance criteria that demystify the demand-to-damage transformation through simple subjective but quantitative guidelines. The need to improve and fine-tune these guidelines is well recognized, but the process is well defined and deterministic.

Response quantities estimated from the analysis of the building model in Step 2 need to be compared to acceptable or allowable limits that are termed "acceptance criteria." These limits can be specified both at the global level in terms of interstory drift limits and at the local level in terms of component demand limits. The limits are a function of performance levels. Hence, to achieve a higher performance level for a given hazard level, the acceptability criteria will be more stringent.

The introduction of acceptance criteria for different performance levels represents a significant departure from traditional seismic design and is obviously the singular step in the overall process that qualifies the term "performance based" in the terminology of structural design. To develop acceptance criteria for a range of element types, it is necessary to have access to information on physically observed damage states during laboratory testing of components and subassemblies. FEMA-356 and ATC-40 provide an initial compilation of such data based on the input of experts and must be regarded as a preliminary guideline for use in the evaluation of existing structures. What is important is that the framework now exists to establish new acceptance limits and engineers have the flexibility to develop their own data using the results of experimental research.

An alternative approach to assessing performance is to describe it in a probabilistic sense given the fact that it is difficult to establish quantitative or deterministic estimates of the consequences of demand limits or damage states. For this reason, FEMA-350 proposes "confidence levels" on the demand estimates that result in a probabilistic description of the performance level. Ongoing work in PEER and MAE attempts to further expand this description using concepts in probability theory. The result is the evolution of cumulative distribution functions or fragility curves that characterize the conditional probability of a decision, damage, or demand variable as a function of the seismic hazard.

To wrap up the introduction to the three-step methodology, it is necessary to return to one of the fundamental objectives that advanced the art of PBSE, namely, the consideration of losses from structural and nonstructural damage. The basic framework of FEMA-356 or ATC-40 does not explicitly link acceptance criteria to expected losses, but it is possible to generate estimates of potential losses given a known damage state. Some of the ongoing research in PBSE (Miranda and Aslani 2003; Krawinkler and Miranda 2004) attempts to address this issue. For example, the PEER methodology explicitly includes decision variables (such as dollar loss, downtime, or other quantifiable economic impact) in its performance-based formulation and will be briefly discussed later in this chapter.

5.3 Performance-Based Methodologies: Deterministic Approach

The precursors to performance-based design in seismic engineering, as suggested earlier in this chapter, can be found in two separate published reports: ATC-40 and FEMA-356. However, both documents arose from the urgent need to develop systematic guidelines for seismic rehabilitation of buildings. The effort to produce ATC-40 was prompted by the stock of pre-1970s cast-in-place concrete buildings in California. The lack of guidelines to assist engineers to rehabilitate these relatively nonductile buildings resulted in the ATC-40 project. Given the narrow focus of ATC-40, it provides a fairly detailed and in-depth look at different perspectives in performance-based design and evaluation. FEMA-356, on the other hand, was a much larger effort and covers all building types: steel, concrete, masonry, wood, and structures incorporating isolation or energy dissipating devices. Though both documents employ a similar format for describing performance objectives, only ATC-40 endorses the capacity spectrum method (CSM) as a basis for evaluating structural performance.

We begin with the common thematic elements in ATC-40 and FEMA-356, namely, the classification of performance levels leading to the specification of performance objectives.

5.3.1 Classifying Performance Levels

Structural and nonstructural performance levels are generally combined to yield the overall building performance level. At the structural level, the following major performance levels are used:

- Immediate occupancy (IO) level requires the building to be safe for unlimited egress, ingress, and occupancy.
- Damage control is described as a performance range from IO to life safety level rather than a specific performance level.
- Life safety (LS) level of performance suggests structural damage without partial or total collapse of structural members, which might pose a risk to life.
- Limited safety, like damage control, is another performance range that lies between LS and the structural stability limit.
- Structural stability level can be viewed as an alternative definition of collapse prevention (CP) wherein the structure is still capable of maintaining gravity loads though structural damage is severe and the risk of falling hazards is high.

Each structural performance level is associated with a damage state that can be observed or quantified. For example, FEMA-356 describes the expected damage at LS performance level for concrete frames as "extensive damage to beams," "hinge formation" in ductile secondary elements with crack widths less than $\frac{1}{8}$ in., and permanent drifts up to 1%. At CP performance level, concrete frames could experience permanent drifts up to 4%, while steel frames may see drift (transient or permanent) magnitudes of 5%. Both FEMA-356 and ATC-40 (which applies to concrete structures only) contain detailed descriptions of damage for different element types (columns, beams, structural walls, etc). It must be reiterated that the damage measures are based on the expected performance of existing buildings. The magnitudes of

TABLE 5.1 Major Performance Levels Recommended in ATC-40 and FEMA-356

Nonstructural performance levels	Structural performance levels		
	I: immediate occupancy	III: life safety	V: structural stability (collapse prevention)
A: operational	I-A: operational	NR	NR
B: immediate occupancy	I-B: immediate occupancy	III-B	NR
C: life safety	I-C	III-C: life safety	V-C
E: not considered	NR	NR	V-E: structural stability

Note: NR, not recommended.

these measures can be adjusted if experimental data are available to support alternative performance ranges.

Nonstructural performance levels include the performance of components such as elevators, piping, fire sprinkler systems, and heating, ventilation, and air conditioning (HVAC) equipment and building contents such as computers, bookshelves, and art objects. Nonstructural performance levels can be defined as follows:

- Operational performance level requires the postearthquake damage state of all nonstructural elements to remain functional and in operation.
- IO performance level allows for minor disruption due to shifting and damage of components, but all nonstructural components are generally in place and functional.
- LS performance level allows for damage to components but does not include failure of items heavy enough to pose a risk of severe injuries or secondary hazards from damage to high-pressure toxic and fire-suppressing piping.
- Reduced hazard is a postearthquake damage state that considers risk to groups of people from falling heavy objects such as cladding and heavy ceilings.
- Finally, it is also possible to include what is called a "not considered" performance level. This performance level is provided to cover those nonstructural elements that have not been evaluated as having an impact on the overall structural response. In fact, it is common for computer models evaluating seismic demands to ignore the presence of nonstructural elements.

When structural and nonstructural performance levels are combined, the resulting building performance level is established. Table 5.1 shows how such a performance level is determined. Note that only a few selected performance levels are displayed in Table 5.1. In the context of Table 5.1, if a building performance level of III-C were selected, this would require LS performance level at both structural and nonstructural elements.

5.3.2 Defining Performance Objectives

To prescribe a set of performance objectives, ATC-40 defines the following three hazard levels: (1) a serviceable earthquake (SE) with a 50% probability of being exceeded in 50 years, (2) a design earthquake (DE) with a 10% probability of being exceeded in 50 years, and (3) a maximum earthquake (ME) with a 5% chance of being exceeded in 50 years. FEMA-356, on the other hand, specifies four hazard levels in which the earthquake event has the following probabilities: 50% in 50 years, 20% in 50 years, 10% in 50 years, and 2% in 50 years.

A performance objective is now defined by selecting a building performance objective (Table 5.1) for a given hazard level. A possible set of combinations is displayed in Table 5.2 for three hazard levels and three performance levels, making a total of nine possible combinations. For example, a possible performance objective would be to achieve an overall building performance level of III-C for a DE event which translates into LS performance for a 10%/50-year earthquake (or *b* in Table 5.2). It is also

TABLE 5.2 Defining Performance Objectives

	Target building performance levels		
Hazard level	Immediate occupancy	Life safety	Structural stability or collapse prevention
10%/50 years	*a*	*b*	*c*
5%/50 years	*d*	*e*	*f*
2%/50 years	*g*	*h*	*i*

possible to create dual or multiple performance objectives by selecting different performance levels and different hazard levels. An example of this would be achieving I-B performance level for a DE event but satisfying III-C for an ME earthquake (this would correspond to *a* and *e* in Table 5.2).

Both ATC-40 and FEMA-356 prescribe a basic safety objective (BSO), which comprises a dual-level performance objective. In ATC-40, this would entail satisfying III-C (LS performance level) for a DE and also achieving V-E (structural stability) for an ME event. In FEMA-356, the BSO criterion requires LS performance for a 10% in 50-year event and CP performance level for a 2% in 50-year earthquake. With reference to Table 5.2, the BSO in ATC-40 is *b* + *f* and in FEMA-356 it is *b* + *i*.

5.3.3 Design Response Spectra and Ground Motions

The selection of a performance objective, as just described, involves the specification of a hazard level. Unless ground motion time histories are used in a dynamic time-history analysis, it is customary to specify the hazard in terms of a response spectrum. The generation of the ground motion hazard spectrum is a function of several parameters, most of which pertain to site characteristics.

Site-specific hazard analysis is recommended for very soft soils that are vulnerable to failure under seismic action such as liquefiable soil, highly sensitive clays, high-plasticity clays with depths exceeding 25 ft and $PI > 75$, and soft- to medium-stiff clays with depths exceeding 120 ft. Soil properties are based on average values for the top 100 ft (30 m) of soil profile. Site-specific studies are also required for special and critical structures located near an active fault. Even if a site-specific hazard spectrum is developed, both FEMA-356 and ATC-40 place limits on the values of site response coefficients C_A and C_V.

In all other cases, an elastic response spectrum as shown in Figure 5.4 can be constructed from known values of C_A and C_V. Typical soil profile types are described in Table 5.3.

5.3.3.1 Generating the Design Spectrum Using ATC-40 Provisions

The site response coefficient C_A is simply the effective peak acceleration (EPA) at the site while the coefficient C_V when divided by the period defines the acceleration in the constant velocity domain. ATC-40 provides the following three options when developing the elastic design spectra:

1. Site specific hazard analysis.
2. Spectral contour maps developed by U.S. Geological Survey (USGS).
3. Site seismic coefficients given in Table 5.4.

If the USGS contour maps (developed for rock sites) are used, then the following modifications are prescribed for buildings on S_B soil sites: $C_A = 0.4S_{MS}$, $C_V = S_{M1}$, where S_{MS} is the spectral acceleration at 0.3 s (short-period range) and S_{M1} is the spectral acceleration at 1.0 s (velocity-sensitive range).

To use Table 5.4, it is first necessary to determine the shaking intensity, which is defined as the product of three quantities: $Z * E * N =$ zone factor × earthquake hazard level × near source factor. Using the shaking intensity value, the corresponding site response coefficients C_A and C_V are obtained for a known soil profile. The response spectrum can now be easily generated as indicated in Figure 5.4.

The potential for liquefaction and landsliding at a site are also important considerations when evaluating a structure for earthquake hazards. ATC-40 provides some guidelines for determining when

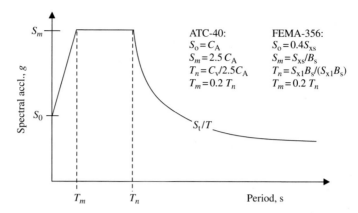

FIGURE 5.4 ATC-40 and FEMA-356 representation of the 5% damped design response spectrum.

TABLE 5.3 Average Soil Properties Used to Establish Soil Profile

Soil profile type	Soil profile name	Shear wave velocity (m/s)	Standard penetration test (blows/30 cm)	Undrained shear strength (kPa)
S_A	Hard rock	>1520	n/a	n/a
S_B	Rock	760–1520	n/a	n/a
S_C	Very dense soil	365–760	> 50	>2
S_D	Stiff soil	180–365	15–150	1–2

Note: n/a, not applicable.

TABLE 5.4 ATC-40 Specifications for Developing Hazard Spectrum

Soil profile		Shaking intensity, *ZEN*					
		0.075	0.150	0.200	0.300	0.400	>0.400
S_B	C_A	0.08	0.15	0.20	0.30	0.40	$1.0 * ZEN$
	C_V	0.08	0.15	0.20	0.30	0.40	$1.0 * ZEN$
S_C	C_A	0.09	0.18	0.24	0.33	0.40	$1.0 * ZEN$
	C_V	0.13	0.25	0.32	0.45	0.56	$1.4 * ZEN$
S_D	C_A	0.12	0.22	0.28	0.36	0.44	$1.1 * ZEN$
	C_V	0.18	0.32	0.40	0.54	0.64	$1.6 * ZEN$

Notes: $Z =$ zone factor (IBC-2000: e.g., Zone $4 = 0.4g$); $E = 0.5$ (SE), 1.0 (DE), and 1.25 (Zone 4) and 1.50 (Zone 3) (ME); $N =$ near source factor (typically, $N = 1$ for faults not capable of producing events with maximum moment magnitude $M > 6.5$ and faults with slip rates less than 2 mm/year). Linear interpolation is permitted for intermediate values.

a more detailed study of these hazards is warranted. In both cases, the expected EPA is used to establish the likelihood of ground failure given the liquefaction potential or the geologic group at the site.

5.3.3.2 FEMA-356 Provisions for Generating the Design Spectrum

National Earthquake Hazards Reduction Program (NEHRP) seismic maps (also available at http://eqhazmaps.usgs.gov) provide peak acceleration plots for 5% damped response spectrum at two characteristic periods (short duration: 0.2 s, long duration: 1 s). These maps are referenced to site class B and must be adjusted for the applicable site class through the following equations:

$$S_{xs} = F_a S_s \tag{5.3}$$

$$S_{x1} = F_v S_1 \tag{5.4}$$

where S_{xs} is the design short-period spectral response parameter, S_{x1} is the design spectral response acceleration parameter at 1 s, F_a and F_v are the site coefficients (provided in FEMA-356), S_s is the short-period spectral response parameter, and S_1 is the spectral response acceleration parameter at 1 s.

The site-specific response spectrum is then generated using the following equations:

$$S_a = (S_{xs}/B_s)(0.4 + 3T/T_n) \quad \text{for } 0 < T \leq T_m \tag{5.5}$$

$$S_a = (S_{xs}/B_s) \qquad\qquad\qquad \text{for } T_m < T \leq T_n \tag{5.6}$$

$$S_a = (S_{x1}/B_1 T) \qquad\qquad\qquad \text{for } T > T_n \tag{5.7}$$

where

$$T_n = (S_{x1}B_s)/(S_{xs}B_1) \tag{5.8}$$

and B_s and B_1 are damping coefficients, quantified in FEMA-356 as a function of effective damping in the system. The resulting response spectrum is conceptually similar to ATC-40 and is shown in Figure 5.4.

5.3.4 Ground Motions

Both ATC-40 and FEMA-356 contain provisions for using acceleration time histories in conjunction with transient response analyses. A minimum of three ground motion sets must be used. The selected ground motions are expected to have magnitude, fault distances, and source mechanisms that are comparable to the ground-shaking hazard at the building site. FEMA-356 requires that the data sets be scaled in such a manner that the average value of the square root of sum of squares (SRSS) spectra (that are generated for each data set) does not fall below 1.4 times the 5% damped spectrum for the DE for periods between $0.2T$ and $1.5T$ (where T is the fundamental period of the structure).

When only three time history sets (one horizontal and one vertical component for two-dimensional [2D] systems and all three components for a 3D model) are used in the evaluation process, the maximum value of each response parameter should be considered. If seven or more records are used, then both FEMA-356 and ATC-40 permit mean response estimates to be considered to assess design acceptability.

ATC-40 actually provides a list of ten earthquake data sets for two scenarios: sites that are farther than 10 km from the source and sites that are within 5 km from the source to be considered near-fault events. The ground motions listed in ATC-40 are free-field motions recorded on firm soil sites with magnitude greater than 5.5 and a PGA exceeding 0.2*g*. A subset of these records is used in the design example presented later in this chapter. It is expected that the listed ground motions be scaled such that the average value of the spectra matches the site response spectrum in the period range of interest (an effective period range near the performance point of the structure — see ensuing discussion on CSM for a definition of the performance point).

5.4 Evaluating Seismic Demands

The estimation of seismic demands requires the development of a mathematical model of the building. The model should incorporate all components that influence the mass, stiffness, and strength of the building, particularly in the inelastic regime of the response. Hence, issues like soil–structure interaction must be carefully evaluated before a decision is made to include or exclude the soil–foundation system in the final model. The model should properly account for gravity loads that comprise dead loads and other permanent fixtures. Consideration of live loads is also necessary if the presence of additional gravity loads is likely to create a situation resulting in adverse seismic response or the shifting of the location of the plastic hinge.

A structural model also includes specification of the expected behavior of all of the elements used to develop the building model. A linear elastic analysis requires only the estimation of the effective stiffness of each element, whereas a nonlinear analysis demands a more concerted effort to establish the expected local behavior of every element in the overall structural model. The guidelines summarized in the next section are partly based on recommendations collectively provided in ATC-40, FEMA-356, and FEMA-350.

5.4.1 Modeling Guidelines

The development of analytical models must account for all possible aspects of behavior while still working within the limitations of the analytical tools being used to carry out the evaluation of seismic demands. Both commercial and noncommercial software is available to assist engineers in estimating seismic demands, and the modeling requirements vary from one tool to the next. Most programs employ simple hinge-based inelastic models with multilinear characterization of force–deformation behavior. In general, it is necessary to model three commonly recognized aspects of cyclic behavior: stiffness degradation, strength deterioration, and pinching response. Hysteretic models that incorporate some or all of these behavioral aspects have been proposed by numerous researchers (Kunnath et al. 1997; Sivaselan and Reinhorn 2000). There is also a large body of work on degrading force–deformation models developed by Japanese researchers, many of which are summarized in a report by Umemura and Takizawa (1982).

A general hysteretic model that incorporates all three elements is shown in Figure 5.5. This basic model, which is implemented in the inelastic damage analysis of RC structures (IDARC) (Kunnath et al. 1992; Kunnath 2004) series of programs, uses several control parameters to establish the rules under which inelastic loading reversals take place. A variety of hysteretic properties can be achieved through the combination of a nonsymmetric trilinear curve and certain control parameters that characterize the shape of the force–deformation loops. For example, α, which can be expressed as a function of the deformation, controls the amount of stiffness loss; ϕ and χ control the initiation and degree of pinching respectively; and the slope s and the change in expected peak strength (M to M^*) control the softening due to system deterioration. A sample simulation of observed behavior using this model is shown in Figure 5.6. The hysteresis curves in this case were obtained from tests of a precast concrete connection with a hybrid combination of mild steel and posttensioning steel (Cheok et al. 1998). However, the specification of hysteretic rules in an actual analysis is rather empirical and should be based on available experimental data. Even so, a parametric study to evaluate the sensitivity of these parameters is necessary.

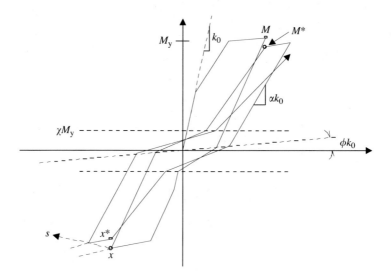

FIGURE 5.5   General-purpose hysteretic model.

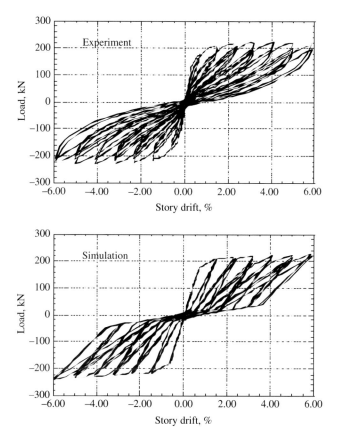

FIGURE 5.6   Simulation of observed hysteretic behavior.

The degree of degeneration is a function of the connection detail. Welded steel moment connections or well-detailed RC members may display stable hysteresis with little degradation of strength in the performance range of the element. Bolted connections or concrete sections with inadequate details may exhibit pinching behavior. Shear dominated members will experience rapid decay in strength after yielding. In general, it is necessary to know in advance the expected behavior of an element when using force–deformation type models. Complete cyclic description of behavior, as illustrated in Figure 5.5 and Figure 5.6, is essential for nonlinear time-history methods only. For nonlinear static approaches, only the force–deformation envelope is needed. FEMA-356 and ATC-40 specify three types of force–deformation envelopes for use with nonlinear static analyses. These are displayed in Figure 5.7. A Type I envelope is meant to represent the expected behavior of a well-detailed ductile connection with a well-defined postyield range (a to b), a region of gradual decay (b to c), and a nonnegligible residual strength that is capable of supporting gravity loads in the regime (c to d). A Type II curve is similar to Type I with the exception that an element modeled as Type II cannot be relied upon to support gravity loads beyond "b." An important characterization in FEMA is the classification of members as being force controlled or deformation controlled. Members modeled as Type I are generally deformation-controlled members while Type II elements can be classified as deformation controlled only if the ductility at peak strength (b/a) is greater than 2.0. A Type III curve is representative of brittle failure and is, therefore, always a force-controlled element.

An additional detail that is outlined in FEMA-350 is the identification and specification of the interstory drift angle and the associated plastic hinge locations. The idea is that frames should be detailed in such a manner that the required interstory drift angle can be accommodated through a combination

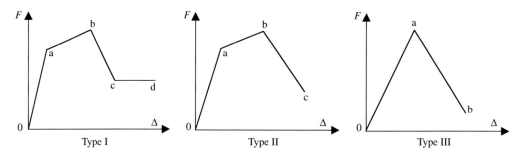

FIGURE 5.7 Envelope patterns recommended in ATC-40 and FEMA-356 for use in nonlinear analysis methods to estimate seismic demands.

of elastic deformation and the development of plastic hinges at known locations within the frame. For example, the use of details such as haunches or reduced beam sections can force yielding in the beam but away from the column face. Figure 5.8 demonstrates this concept. The interstory drift angle, as specified in FEMA-350, is equal to the sum of the plastic drift and that portion of the elastic interstory drift resulting from flexural deformation of the individual members. Interstory drift resulting from axial deformations of columns is not included.

5.4.1.1 *P*–Δ Effects

The consequence of secondary moments resulting from excessive lateral displacements can have a major impact on the performance of a building in the postyield range of the response. *P*–Δ effects alter the postyield stiffness of a structure, and with increasing deformation demands the net effect of *P*–Δ softening can lead to rapid deterioration and collapse of the system. FEMA-356 recognizes both static and dynamic *P*–Δ effects. For linear procedures, it is required to evaluate a stability coefficient given by

$$\theta_i = \frac{P_i \delta_i}{V_i h_i} \tag{5.9}$$

where P_i is the weight of structure acting on story level i, V_i is the total lateral shear force acting on story level i, h_i is the height of story level i, and δ_i is the lateral drift of story i.

If the stability coefficient is less than 0.1 in all stories, then *P*–Δ effects can be ignored in the analysis. For values of the coefficient between 0.1 and 0.33, the seismic effects are to be magnified by a factor $1/(1 - \theta_i)$, and for values in excess of 0.33, the system is considered unstable and should be redesigned.

Even if nonlinear procedures are used, a magnification factor (C_3) is used in FEMA-356 for static methods because dynamic *P*–Δ effects are recognized as being different from static effects. Static *P*–Δ should be incorporated in pushover analysis, while both static and dynamic *P*–Δ can be automatically incorporated in nonlinear dynamic procedures.

5.4.1.2 Other Effects and Considerations

In general, it is expected that a building be modeled as a 3D assembly of elements. However, most structures tend to be fairly regular with assumptions of symmetry commonly employed in routine analysis. Some modeling issues that must be resolved before finalizing a building model for use in demand prediction include the following:

- The assumption of rigid diaphragms, which allows axial deformations in floor beams to be ignored and simplifies the resulting system of equations that need to be solved.
- Any out-of-plane offsets in those vertical elements that resist lateral forces must be accounted for when estimating demands on floor diaphragms.

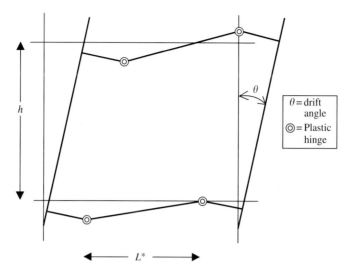

FIGURE 5.8   Definition of interstory drift angle.

- Consideration of actual and accidental horizontal torsion: FEMA-356 lists specific criteria for considering the effects of horizontal torsion.

Normally, it is adequate to consider seismic input in each principal direction of the building only. However, for irregular buildings, concurrent seismic effects in two orthogonal directions must be accounted for by combining 100% of the design effects in one direction with 30% of the design effects in the orthogonal direction. Similarly, the need to consider vertical seismic input arises only in special circumstances such as cantilevered or prestressed elements and those components in which gravity load demands are excessive.

Other considerations not discussed in this section include

- Foundation and SFSI interaction effects.
- Variability in ground motion.
- Modeling of secondary and nonstructural components.
- Introduction and modeling of seismic protection devices such as isolators, dampers, and other energy dissipating devices.

A description of the details that take into consideration the consequence of the above effects is beyond the scope of this chapter.

5.4.2 Methods of Analysis

The estimation of demands can be accomplished using a variety of available procedures. In this section, discussion will be limited to a generic set of evaluation methods that best lend themselves to an assessment within a performance-based framework. The primary objective is to determine forces and deformations both at the global and at the local level when the structure is subjected to seismic loads that characterize the hazard at the building site. As such, there are four possible methods to analyze a mathematical model of a building structure. They may be classified into two broad categories depending on the treatment of the response or the treatment of the loads. The former category results in the distinction between linear and nonlinear methods of analysis, while the latter distinguishes static and dynamic application of the seismic loads.

Linear static and dynamic procedures (LSP and LDP). These procedures are recommended for regular buildings where issues such as torsion and high-mode effects are negligible. The expectation is that the computed displacements using linear equivalent elastic stiffness are approximately equal to the actual displacements that may occur inelastically under the design loads. FEMA-356 lists specific criteria to limit the use of such procedures. For example, it states that linear methods should not be used if the demand to capacity ratio (also called *m*-factors) for any element in the structure exceeds 2.0. Even if *m*-factors exceed 2.0, linear procedures may still be used for regular buildings. Section 2.4.1.1 of FEMA-356 outlines criteria for in-plane and out-of-plane discontinuities in the lateral force resisting system and also for irregularities arising from weak stories or torsional strength. Linear procedures are not recommended in ATC-40.

Nonlinear static and dynamic procedures (NSP and NDP). Nonlinear procedures are generally applicable for all buildings with the exception that NSP is limited to buildings where high-mode effects are small. Again, the FEMA document has explicit guidelines to determine if higher modes play an important role in the response: high-mode effects are deemed significant if the shear in any story resulting from a modal analysis considering modes required to obtain 90% mass participation exceeds the corresponding story shear considering first-mode response only by a factor of 1.3.

The LSP is the simplest procedure described in FEMA-356. LSP is an equivalent lateral load analysis procedure that attempts to represent the seismic loading as a static force. The reason for retaining a linear static approach in a performance-based guideline is debatable, but the motivation stems from the simplicity of the method and the current familiarity among engineers with force-based design. It is also felt that such an approach may still be valid for a large group of regular structures. The total lateral load applied to the structure is calculated from

$$V = C_1 C_2 C_3 C_m S_a W \tag{5.10}$$

where V is the total base shear, C_1, C_2, and C_3 are modification factors to account for the effects of inelasticity, system degradation, and P–Δ effects, respectively, C_m is an effective mass factor to account for high-mode participation, and S_a is the spectral acceleration at the fundamental period of the structure. The seismic weight of the structure, W, includes 25% of live load in addition to dead loads. The fundamental period, T, of the structure can be computed using an eigenvalue analysis or through an empirical equation, given by the following expression:

$$T = C_t (H_n)^\beta \tag{5.11}$$

where H_n is the height of the building measured from base to roof, and coefficient C_t and power β depend on the structural system. The fundamental period is a critical parameter in the evaluation procedure for both static procedures (LSP and NSP), since they can change the applied lateral force in the linear procedure or alter the target displacement in the nonlinear procedure (discussed in Section 5.5.2). The total shear computed in Equation 5.10 is distributed over the building height as follows:

$$F_x = \frac{w_x h_x^k}{\sum w_i h_i^k} V \tag{5.12}$$

where w_i is the portion of weight at floor level i, w_x is the portion of weight at floor level x, H_i is the height from base to floor level i, H_x is the height from base to floor level x, and $k = 1.0$ for $T < 0.5$ s and 2.0 for $T \geq 2.5$ s.

The summation in the denominator of Equation 5.12 is carried over all stories, and the coefficient k in the period range from 0.5 and 2.5 s is established through linear interpolation. Note that a reduction factor is not applied to the computed base shear. Hence, the demands will exceed the strength capacity of those components that are expected to yield under the design earthquake loads. The resulting demand-to-capacity ratios are evaluated at the component level.

The LDP involves a linear elastic dynamic analysis using a response spectrum approach or direct time-history evaluation to determine the building response. Linear elastic stiffness properties of elements and equivalent viscous damping at or near the yield level are to be used in the analysis. An unmodified elastic spectrum is utilized in a response spectrum analysis. Peak forces and deformations are estimated using an SRSS or complete quadratic combination (CQC) of modal quantities. For time-history methods, if less than seven earthquake records are used, the maximum response quantities are to be used in the evaluation procedure. For seven or more records, an average value of the response is adequate. As mentioned previously, FEMA-356 requires that each record be scaled such that the average value of the SRSS of the 5% damped spectrum does not fall below 1.4 times the design spectrum value (also constructed for 5% damping) between $1.2T$ and $1.5T$ (where T is the fundamental building period).

The NSP or pushover analysis takes into consideration material nonlinearities. The concept gained prominence after its introduction in the CSM by Freeman (1978). The nonlinear static method is also called the displacement coefficient method in that the expected demand (target displacement) of a single degree-of-freedom (SDOF) system is modified by a set of coefficients to approximate the multiple DOF response. The procedure involves selecting a control node (typically the center of mass at the roof level of the building) and subjecting the structural model to monotonically increasing lateral loads till a target displacement is reached. The lateral load applied to the building to achieve the control node target displacement can be distributed in several ways. The choice of a lateral load pattern can have a significant influence on the calculated response. FEMA-356 suggests that at least two patterns be used: a uniform pattern that results in lateral forces proportional to the total mass of each floor level and a modal pattern, which should be selected from the following two options: (1) when more than 75% of the total mass participates in the fundamental mode, a lateral load pattern represented by Equation 5.12 is used and (2) a lateral load pattern that is proportional to the story shear distribution, calculated by combining modal responses from a response spectrum analysis wherein at least 90% modal mass participation is incorporated, is used. The target displacement is established from the following equation:

$$\delta_t = C_0 C_1 C_2 C_3 S_a \left[\frac{T_e^2}{4\pi^2} \right] g \qquad (5.13)$$

where δ_t is the target displacement (note that response quantities are determined when the roof displacement reaches this target displacement), C_0 is a modification factor relating spectral displacement estimated for an equivalent SDOF system to the likely roof displacement of a multistory structure, and the remaining coefficients C_1 to C_3 have the same meaning as previously defined for linear procedures. The effective fundamental period T_e is determined from

$$T_e = T_i (K_i / K_e)^{0.5} \qquad (5.14)$$

where T_i is the elastic fundamental period, K_i is the elastic lateral stiffness, and K_e is the secant stiffness at 60% of the yield strength of the building.

Pushover methods have been the subject of numerous studies following numerous questions about the validity of using static procedures to predict dynamic demands. Modified and advanced pushover techniques are available in the literature and are discussed later in this chapter following the design example.

The NDP refers to a complete nonlinear dynamic time history analysis and is generally regarded as the most accurate method of demand evaluation. A complete nonlinear description, in terms of force–deformation behavior at a cross-section (either using hysteretic models or from direct integration of stresses derived from nonlinear constitutive relationships), of all elements expected to experience inelastic deformations should be developed. No fewer than three ground motion sets (comprising of two horizontal components and a vertical component, if necessary) need to be considered. The selected time histories should have magnitude, fault distance, and source mechanisms that are equivalent to the design ground motion at the site. More detailed criteria are specified in FEMA-356 if simulated motions are used. Similar to the linear dynamic procedure, mean response quantities can be used if seven or more records are utilized in the simulations, and peak response quantities should be used if fewer than seven ground motions are used in the evaluation.

5.5 Assessing Performance: The Development of Acceptance Criteria

The third and final step in a general PBD process is to determine if the demands computed in step 2 using one of the four methods described in the previous section are within the acceptance criteria that satisfy the performance level defined as part of the overall performance objective in step 1 of the procedure. The current thinking in FEMA-356 is that the performance of a component in the system is critical to the overall performance of the building. Consequently, acceptance criteria are specified at the element level. The resulting responses from the analysis of the structural model of the building system are classified as force-controlled (zero or limited ductility) or deformation-controlled actions. It is also necessary to distinguish primary and secondary components.

5.5.1 Linear Procedures

For linear procedures (LSP and LDP), deformation-controlled design actions should include the combined effects of earthquake and gravity loads as follows:

$$Q_{UD} = Q_G \pm Q_E \tag{5.15}$$

where Q_{UD} is the deformation-controlled design action, Q_G is the action due to gravity loads, and Q_E is the action due to earthquake forces.

In the case of force-controlled actions (defined as actions in components where nonlinear deformations are not permitted or are limited), the following expression is used in FEMA-356:

$$Q_{UF} = Q_G \pm \frac{Q_E}{C_1 C_2 C_3 J} \tag{5.16}$$

where Q_{UF} is the force-controlled design action, Q_G is the action due to gravity loads, Q_E is the action due to earthquake forces, C_i are the amplification coefficients introduced in the previous section, and J is the smallest demand-to-capacity ratio of components delivering forces to the component (≥ 1); alternatively, $J = 2$ in high seismic zones, 1.5 in moderate seismic zones, and 1.0 in zones, of low seismicity.

Note that both the positive and the negative signs that appear in Equations 5.15 and 5.16 should be considered when determining the design actions, and the worst-case scenario should be used in the evaluation. Following the analysis using either an LSP or an LDP, the demand-to-capacity ratio (or *m*-factors) for each component is established as follows:

$$m = \frac{Q_{UD}}{\kappa Q_{CE}} \tag{5.17}$$

where Q_{CE} is the expected capacity of the component (typically the yield capacity of the section considering possible interaction with other actions) and κ is the knowledge factor that accounts for the uncertainty in estimating the material properties of the section (this factor is unity for new designs).

The demand-to-capacity ratio for force-controlled elements is limited to unity: elements with limited ductility (less than 2.0) are also classified as force-controlled elements within the scope of FEMA-356 guidelines. For deformation-controlled elements (those elements where inelastic action is expected and permitted), a set of acceptance (or allowable) limits need to be developed for different components. Typically, such values come from an evaluation of experimental data supplemented by analytical studies. FEMA-356 provides an initial set of numbers that are recommended for existing buildings. However, the FEMA tables also contain recommended values for well-detailed sections that can form the basis of

TABLE 5.5   Typical Acceptance Criteria for Primary Components in a Fully Restrained Steel Frame Building

	m-Factors for linear procedures			Plastic rotation angle for nonlinear methods		
Component	IO	LS	CP	IO	LS	CP
Beams — flexure						
$\dfrac{b_f}{2t_f} \le \dfrac{52}{\sqrt{f_{ye}}}$ and $\dfrac{h}{t_w} \le \dfrac{418}{\sqrt{f_{ye}}}$	2	4	6	θ_y	$6\theta_y$	$8\theta_y$
Columns — flexure $P/P_{CL} < 0.20$						
$\dfrac{b_f}{2t_f} \le \dfrac{52}{\sqrt{f_{ye}}}$ and $\dfrac{h}{t_w} \le \dfrac{300}{\sqrt{f_{ye}}}$	2	4	6	θ_y	$6\theta_y$	$8\theta_y$

TABLE 5.6   Typical Acceptance Criteria for Primary Components in a Reinforced Concrete Frame Building

			m-factors			Plastic rotation angle		
			IO	LS	CP	IO	LS	CP
Beams (flexure controlled)								
$\dfrac{\rho - \rho'}{\rho_{bal}}$	Transverse reinforcement	$\dfrac{V}{b_w d \sqrt{f_c'}}$						
≤ 0.0	C	≤ 3	3	6	7	0.010	0.020	0.025
≤ 0.0	NC	≤ 3	2	3	4	0.005	0.010	0.020
Columns (flexure controlled)								
$\dfrac{P}{A_g f_c'}$	Transverse reinforcement	$\dfrac{V}{b_w d \sqrt{f_c'}}$						
≤ 0.1	C	≤ 3	2	3	4	0.005	0.015	0.020

acceptance criteria for new design. A sample of typical *m*-factors for both steel and concrete members is shown in Table 5.5 and Table 5.6.

5.5.2 Nonlinear Procedures

Acceptance criteria for nonlinear procedures, both in ATC-40 and in FEMA-356 are based on peak demands without explicit consideration of cumulative effects resulting from cyclic effects. To some extent, cyclic degrading effects can be implicitly included in the specification of the backbone envelope. To process the results of a nonlinear analysis, it is essential to define response measures that form the basis of acceptance criteria. Currently, the response measure of choice is the rotation demand at the plastic hinge. If a concentrated hinge model is used in the analysis, this is a relatively direct response quantity. However, if a spread-plastic model or a fiber-section model is used to represent material nonlinearities in the element, then the computation of plastic rotations is not straightforward. Either a plastic hinge length needs to be defined to integrate curvature estimates or alternative procedures need to be developed to estimate the plastic rotation demand. FEMA-356 and FEMA-350 define deformations in terms of chord rotations as shown in Figure 5.9.

If the FEMA definition of element rotation is used, then two possibilities exist for defining the plastic rotation. The first approach is to track the moment–rotation behavior at every connection and then compute the difference between the peak rotation and the recoverable rotation (see Figure 5.10). In a nonlinear time-history analysis, the recoverable rotation is more completely defined because member behavior is explicitly defined using constitutive material models or hysteresis models. For nonlinear static procedures, the recoverable rotation may be estimated using the initial stiffness path as the unloading path. Alternatively, the yield rotation can be predetermined using conventional concepts in structural mechanics. For example, FEMA suggests the following expression for steel frame members:

$$\theta_y = \frac{Z f_{ye} L}{6EI}\left(1 - \frac{P}{P_{ye}}\right) \qquad (5.18)$$

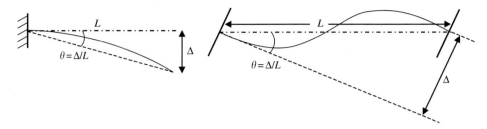

FIGURE 5.9   Definition of chord rotations to be used in calculating plastic rotation demands.

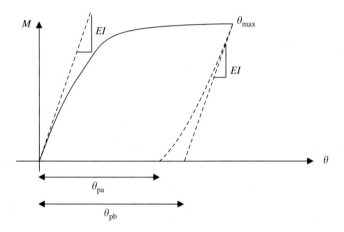

FIGURE 5.10   Defining plastic rotations for seismic loading.

where Z is the plastic section modulus, f_{ye} is the yield stress of the material, L is the length of the member, EI is the flexural rigidity, P is the axial force in member (may be taken as zero for beams), and P_{ye} is the expected axial capacity ($A_g f_{ye}$).

However, Equation 5.18 is derived on the assumption that the inflection point occurs at the midpoint of the element. If this is not the case, then the yield rotation should be computed using the following expression:

$$\theta_y = \frac{M_y^2 \ell}{3(M_y + M_2)EI} \tag{5.19}$$

where M_2 is the moment at one end of the member ($M_2 < M_y$), ℓ is the length of member, E is the elastic modulus, I is the moment of inertia, and M_y is the yield moment of the section. In FEMA-356, M_2 is assumed to be equal to M_y. This assumption is reasonably true for beams in moment frame structures but is generally not valid for columns.

In the case of RC section, the determination of yield rotation presents significantly greater challenges. For purely flexural elements, it may be adequate to determine the yield moment and then assume an equivalent flexural rigidity so as to establish the yield curvature. The yield rotation is then calculated by assuming a plastic hinge length. There are widely accepted guidelines for assigning effective stiffness values to RC components. FEMA-356, for example, recommends the following: $0.5EI_g$ for beams and columns with gravity compressive loads $<0.3A_g f_c'$, $0.7EI_g$ for columns with gravity compressive loads $>0.5A_g f_c'$, $0.8\ EI_g$ for uncracked shear walls, and $0.5EI_g$ for walls with visible cracking. The interaction of flexure and axial loads must be considered in developing nonlinear modeling parameters. The interaction of shear and flexure (in the presence of axial forces) is an issue that is still the subject of ongoing research.

Typical limits on plastic rotation demands for different performance levels, as specified in FEMA-356, are also shown in Table 5.5 and Table 5.6 for steel and RC frame buildings.

Ultimately, the intent of the above discussion is to make readers aware of the possible complexities in modeling and evaluating the expected demands at various connections and elements in the structural system. While simple representations of nonlinear behavior lead to more comprehensible results, it is necessary to gain confidence in the computed demands. The desired confidence can be achieved with sensitivity studies on the impact of uncertain variables on the computed demand estimates. For this reason, it is not surprising that FEMA-350 takes a probabilistic approach to PBD. Similarly, ongoing work at PEER evolves around a probabilistic format. Both these methodologies are discussed later in this chapter.

5.5.3 ATC-40 and Capacity Spectrum Procedures

ATC-40 advocates the use of the CSM to evaluate the overall adequacy of design of a structural system. The term "capacity spectrum" refers to an altered form of the capacity or pushover curve for the building. As discussed earlier, a pushover curve provides a representation of both the displacement and the force capacity of a building in terms of roof drift and base shear, respectively. The quantities computed to develop the pushover curve should be recast into spectral displacements and spectral accelerations using simple concepts in structural dynamics. Such a format is conventionally referred to as the acceleration–displacement response spectrum or (ADRS) format (Mahaney et al. 1993). The ATC-40 methodology involves a simple conceptual procedure wherein the reformatted capacity curve is compared to the seismic demand curve, which is also expressed in a similar format. In the context of ATC-40 and capacity spectrum procedures, it should be noted that Reinhorn (1997) proposes an interesting variation on CSM to evaluate the seismic response of buildings. Recently, Fajfar (1999) advanced CSM to incorporate the so-called N2 method into the formulation. These and other related developments on CSM are beyond the scope of this chapter.

5.5.3.1 Determining Capacity

The capacity of a structure is a function of the capacity of its individual components and the interaction of its elements. The preferred method of choice in ATC-40 to establish system capacity is the nonlinear static or pushover procedure in which a mathematical model of the structure is subjected to a lateral force incrementally or iteratively till a capacity curve is obtained. Unlike the pushover procedure in FEMA-356 wherein the lateral load is applied until a target displacement is obtained, in the CSM, the model is pushed till a stability limit is reached. The objective in CSM is to determine the "performance" point of the structure that identifies the demand corresponding to the hazard at the site specified in terms of a response spectrum.

The capacity curve is essentially a pushover curve, and consequently, the NSP procedure of FEMA-356 can be used to develop the curve. If the nonlinear behavior of each element is modeled explicitly, then the resulting base shear versus roof displacement plot is the required capacity curve. However, with the realization that many engineers may only have access to a linear analysis program, ATC-40 also recommends a simplified procedure. In the simplified approach, an elastic analysis is used to approximate the inelastic behavior of the building. The procedure to accomplish this is as follows: a lateral load is applied on the structure, and the magnitude of the lateral load is increased till some element (or a group of elements) reaches approximately 90% of its capacity. At this stage, the analysis is stopped, and the stiffness properties of these elements are set to a very small value or the elements are removed from the model. The lateral loads are reapplied incrementally as before till another set of elements approach their strength capacities. The base shear and roof displacement are recorded at the end of each phase of the analysis, and cumulative values represent points on the capacity curve. The process continues till a stability limit is reached. Figure 5.11a displays the process of combining the results of each phase of analysis. Figure 5.11b shows the same capacity curve, identified in the diagram as Curve #1.

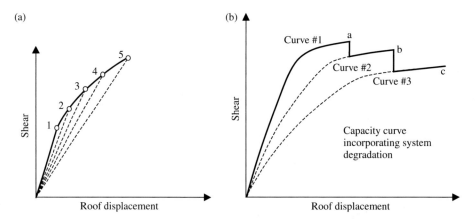

FIGURE 5.11 Generating the capacity curve using linear solution methods (ATC-40).

ATC-40 also recommends changing the lateral load pattern to reflect the displaced shape of the building at the end of each phase of the analysis.

It is also possible to consider degrading effects in the simplified procedure described above by rein-itializing the analysis at the end of each phase and plotting a new capacity curve. Each new curve begins with a degraded system that accounts for the state of elements at the end of each phase. Figure 5.11b provides a conceptual view of three such capacity curves. The final capacity curve is drawn by connecting the final point on the previous curve to a new point on the next capacity curve at the next displacement increment so as to produce a saw-tooth response as shown in Figure 5.11b.

5.5.3.1.1 Conversion to ADRS Format

The capacity curve obtained either by the simplified process described in the previous section or by a detailed nonlinear analysis technique is transformed into ADRS format. The following relationships provide the conversion to ADRS format:

$$S_{a,n} = \alpha_n(V_n/W) \text{ (expressed in units of } g) \tag{5.20}$$

$$S_{a,n} = \frac{\Delta_{c,n}}{\beta_n \Phi_{c,n}} \tag{5.21}$$

$$\alpha_n = \frac{\sum_{i=1}^{N} w_i \Phi_{i,n}^2}{\left(\sum_{i=1}^{N} w_i \Phi_{i,n}\right)^2} \tag{5.22}$$

$$\beta_n = \frac{\sum_{i=1}^{N} w_i \Phi_{i,n}}{\sum_{i=1}^{N} w_i \Phi_{i,n}^2} \tag{5.23}$$

where $S_{a,n}$ is the spectral acceleration for mode n, $S_{d,n}$ is the spectral displacement for mode n, V_n is the base shear for mode n, W is the seismic weight of the building, w_i is the seismic weight of the floor at level i, $\phi_{i,n}$ is the modal amplitude at level i for mode n, and $\Delta_{c,n}$ and $\Phi_{c,n}$ are the displacement and modal amplitude of control node for mode n.

The theoretical basis for making the conversion is outlined in Appendix A.

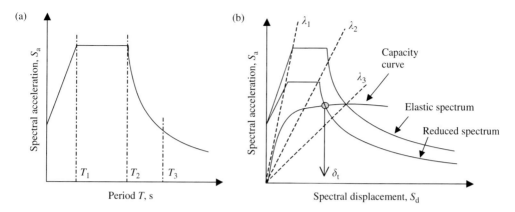

FIGURE 5.12 Generating the ADRS demand and capacity spectrum.

5.5.3.2 Determining Demand

The next step in CSM is to convert the response spectrum (which represents the demand side of the equation) into ADRS format as well, thereby permitting a comparison of demand versus capacity. The design response spectrum is usually expressed in terms of spectral acceleration and period as shown in Figure 5.12a. Since a response spectrum results from the analysis of an SDOF system, the following relationships between pseudospectral acceleration and spectral displacement can be used:

$$S_d = \frac{S_a T^2}{(2\pi)^2} \tag{5.24}$$

$$\frac{S_a}{S_d} = \lambda = \left(\frac{2\pi}{T}\right)^2 \tag{5.25}$$

Hence, it is possible to convert the spectral coordinates for every value of the period T_i into spectral displacements. The resulting curve is in ADRS format and the radial lines shown in Figure 5.12b represent λ values for corresponding T_i values. The capacity curve obtained in the previous step can now be superimposed on the demand curve. If the elastic design spectrum is used to create the demand spectrum, the overlay is valid only if the structural response is also elastic. Hence, the next step in the process is to reduce the elastic response spectrum to an inelastic spectrum using the concept of equivalent damping. Using the fundamental principles of structural mechanics, the equivalent damping ζ_d associated with dissipated energy during inelastic response is given by

$$\zeta_d = \frac{1}{(\Omega/\omega)} \frac{1}{4\pi} \frac{E_D}{E_S} \tag{5.26}$$

where (Ω/ω) is the ratio of the forcing frequency to the natural frequency of the system, E_D is the energy dissipated through hysteretic behavior, and E_S is the strain energy at the maximum displacement. If it is assumed that the peak response is associated with the resonant frequency, then the ratio $\Omega/\omega = 1.0$.

The ATC-40 methodology for estimating the equivalent viscous damping is derived for a bilinear capacity curve, therefore, it is necessary to transform the capacity curve into bilinear form. Figure 5.13 shows a bilinear capacity curve and the energy dissipated in a single cycle given the maximum displacement demand and the corresponding spectral acceleration (S_{dm}, S_{am}). The equivalent damping corresponding to the dissipated energy can be computed using Equation 5.26 and reduced to the following form:

$$\zeta_d = \frac{0.637(S_{ay} S_{dm} - S_{dy} S_{am})}{S_{am} S_{dm}} \tag{5.27}$$

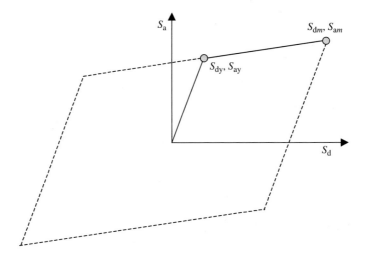

FIGURE 5.13 Bilinear representation of capacity curve and peak demand parameters.

Note that the elastic design spectrum already incorporates 5% damping, hence the equivalent damping (Equation 5.27) from inelastic behavior must be added to the elastic viscous damping. For behavior other than bilinear hysteresis, a modification factor κ is introduced. The final damping value incorporating elastic damping, equivalent inelastic damping, and general hysteretic behavior is given by

$$\zeta_{eq} = \kappa\zeta_d + 0.05 \tag{5.28}$$

Finally, the elastic spectrum is transformed into a reduced spectrum for the damping ratio given by Equation 5.28. ATC-40 provides the following spectral reduction factors, which are derived using the well-known Newmark–Hall relationships:

$$\mathrm{SR_A} = \frac{3.21 - 0.68 \ln(100\zeta_{eq})}{2.12} \tag{5.29}$$

$$\mathrm{SR_V} = \frac{2.31 - 0.41 \ln(100\zeta_{eq})}{1.65} \tag{5.30}$$

where $\mathrm{SR_A}$ is the reduction factor in the constant acceleration region of the spectrum and $\mathrm{SR_V}$ is the reduction factor in the constant velocity region. There are imposed limits on the above reduction factors depending on the expected shape of the hysteresis loops.

5.5.3.2.1 *Performance Point*
The CSM methodology attempts to predict the expected peak displacement given a demand response spectrum. A trial displacement value δ_t (Figure 5.12b) and the corresponding spectral acceleration is selected on the capacity curve. The equivalent viscous damping associated with these spectral magnitudes can now be estimated using Equation 5.27. The 5% damped elastic spectrum is transformed into a reduced inelastic spectrum using Equations 5.20 and 5.21. The demand and capacity curves in ADRS format are overlaid. If the trial displacement is within 5% of the displacement at the intersection of the demand and capacity curves, the performance point has been located. Figure 5.12b shows a case where the trial displacement is the performance point.

5.5.3.3 Performance Assessment
Once the performance point has been established, the acceptability of the design can be assessed by comparing the demand to the acceptance criteria. ATC-40 global acceptance criteria for different performance levels are listed in Table 5.7. Component acceptance criteria are essentially similar to FEMA-356 specifications.

TABLE 5.7 Global Acceptance Criteria for Use with ATC-40 CSM

	Performance level		
Parameter	IO	Damage control	LS
Maximum interstory drift	0.010	0.01–0.02	0.02
Maximum inelastic drift	0.005	0.005–0.015	No limit

5.6 Performance-Based Methodologies: Probabilistic Approaches

5.6.1 The SAC-FEMA Project and FEMA-350

The discovery of numerous brittle fractures of beam-to-column connections in fully restrained steel moment frames following the 1994 Northridge Earthquake dramatically altered a widely held conviction that welded-steel moment-frame buildings are among the most ductile systems contained in the building code. The observed damage was not limited to any particular class of steel structures: varying degrees of brittle fracture were reported in buildings ranging from 1 to 26 stories in height and in a range of ages from 30-year-old buildings to structures being constructed at the time of the earthquake.

The need to reevaluate both existing code provisions and construction practice in welded-steel frames led to the organization of the SAC Joint Venture in 1994, which eventually produced FEMA-350 (2000), a performance-based guideline for the design of new steel structures. The basic procedure described in FEMA-350 is similar to that in FEMA-356 with the major exception that the eventual acceptance criteria assume a probabilistic format. Additionally, it also provides specific qualification criteria for various types of connections that are beyond the scope of this chapter.

The conceptual basis of the FEMA-350 evaluation methodology, outlined in a paper by Cornell et al. (2002), can be expressed in probabilistic terms as follows:

$$P(D > PL) = \int P_{D>PL}(x)h(x)\,\mathrm{d}x \tag{5.31}$$

$P(D > PL)$ is the probability that the damage exceeds a specified performance level within a given duration (say, t years). $P_{D>PL}(x)$ is the probability that the damage exceeds a specified performance level, as a function of x, given that the ground motion intensity is level x. $h(x)\,\mathrm{d}x$ is the probability of experiencing a ground motion of level x to $(x + \mathrm{d}x)$ in a duration of t years.

The steps involved in evaluating the performance of a building using the above concept are summarized here:

1. As with any performance-based process, the performance objective consisting of a hazard level and a performance level is first defined. FEMA-350 lists only two primary hazard levels based on current specifications in FEMA-302: a maximum considered earthquake (MCE), which corresponds to a 2% probability of being exceeded in 50 years, and a DE, which is specified as having an intensity equal to two thirds of MCE. Recall that the DE in ATC-40 was defined as an event with a 10% chance of being exceeded in 50 years. More importantly, only two performance levels are considered: IO and CP. To this extent, FEMA-350 is a limited application of PBD concepts. However, this is an acknowledgment of the lack of available data to classify additional performance limit states.
2. A hazard spectrum or a set of ground motions that meet the criteria specified in the performance objective is developed. The procedure for generating a site-specific response spectrum (similar to

Figure 5.3) is identical to FEMA-356. For probabilistic-based assessment, the ground motion intensity is expressed in terms of the 5% spectral acceleration (S_a) magnitude at the fundamental period of the building. Procedures for determining S_a for other exceedance probabilities are outlined in FEMA-356.

3. The next step in the process is to establish seismic demands that are established from a static or dynamic analysis of a mathematical model of the building — a process not different from that described earlier in this chapter. However, FEMA-350 clearly defines the scope of each analytical method to carry out the task of demand estimation. A table listing the applicability of each method (LSP, NSP, LDP, and NDP) is provided for guidance. For example, if IO performance level is being evaluated and the fundamental period (T) of the structure is less than 3.5 times the characteristic period of the design response spectrum (T_n), then any of the four methods of analysis is valid, but if $T > 3.5T_n$ then static methods are not permitted. Similarly, if CP performance level is being considered for a regular structure with $T < 3.5T_n$ and strong column conditions (sum of column moment capacity is greater than sum of beam moment capacity at connection) exist throughout the frame, then all four methods of analysis are permitted while only nonlinear methods are allowed for irregular structures. Additionally, the demand values of interest are somewhat different from those used in FEMA-356. The following comprise the demand quantities of interest in FEMA-350: interstory drift angles, column compression, and column splice tensile demands.

4. Following the determination of the global and local demand measures, median estimates of structural capacity are established. These capacity parameters include system level measures such as interstory drift capacity and component level measures such as column compressive capacity. FEMA-350 recommends using an incremental dynamic analysis (IDA) to estimate interstory drift capacity, while statistical analyses of available experimental data, as described later in this section, can be utilized for component measures of capacity.

5. Finally, a demand and resistance factor design format is used to determine the confidence level associated with the probability that a building will have less than a specified exceedance probability for a desired performance level. The confidence index is determined through evaluation of the factored-demand-to-capacity ratio given by the equation:

$$\lambda = \frac{\gamma \gamma_a D}{\phi C} \qquad (5.32)$$

where C is the estimated capacity of the structure for the primary demand measures, namely, interstory drift, column compression, and column splice tensile demand; D is the estimated demand for the structure, obtained from the structural analysis of a building model for the corresponding demand measure; γ is a factor that accounts for the variability inherent in the prediction of demand related to structural modeling assumptions and specification of ground motion parameters; γ_a is a factor that accounts for the bias and uncertainty inherent in the method of analysis used to estimate demands; and ϕ is a resistance factor that accounts for the uncertainty and variability inherent in the prediction of structural capacity as a function of ground-shaking intensity.

The confidence level of the performance evaluation is then back-calculated from the following equation:

$$\lambda_L = \exp(-b\beta_{UT}[K_x - 0.5k\beta_{UT}]) \qquad (5.33)$$

where b is a coefficient relating an incremental change in the seismic demand measure to an incremental change in the intensity measure of the ground motion, determined from the demand hazard curve; β_{UT} is a parameter to account for uncertainties in demand and capacity calculated from $\sqrt{(\sum \beta_{ui}^2)}$, where β_{ui} are the standard deviations of the natural logarithms of the variations in demand and capacity, outlined in subsequent sections; k is the slope of the hazard curve (in ln–ln coordinates) at the hazard level of interest; and

K_x is the standard Gaussian variate associated with probability x of not being exceeded as a function of standard deviation measures (obtained readily from standard probability tables).

FEMA-350 includes a table that provides the solution of the above equation for various values of k, λ, and β_{UT}.

The overall confidence level is controlled by the lowest λ value for each of the demand measures (interstory drift, column compression, and column splice tensile demand). The various parameters needed to evaluate the confidence measure are summarized in the following.

5.6.1.1 Slope of the Hazard Curve

A plot of the probability of exceedance (or annual rate of exceedance) of a spectral amplitude versus the spectral value (at a period typically corresponding to the fundamental period of the structure), plotted on a log–log scale, is represented by the following functional form in the hazard range of interest:

$$v(S_{\text{a}}) = k_0(S_{\text{a}})^{-k} \tag{5.34}$$

where k is the slope of the hazard curve and $v(S_{\text{a}})$ is the probability of an event having a spectral amplitude greater than S_{a}. Hazard curves can be developed from seismic hazard maps provided by the USGS and are assessable at http://eqhazmaps.usgs.gov.

5.6.1.2 Demand Variability Factor (γ)

This factor accounts for the variability in the prediction of demand measures such as interstory drift ratio and column splice tension demands. This is determined from the following expression:

$$\gamma = \exp\left(\frac{k}{2b}\beta_{\text{DR}}^2\right) \tag{5.35}$$

where k and b are the same parameters described in Equation 5.33, and β_{DR} is the standard deviation of the natural logarithms of the selected response measure developed from nonlinear time-history analyses of a mathematical model of the building subjected to a suite of ground motions, all of which are scaled to match the 5% damped spectral response acceleration of the hazard spectrum.

5.6.1.3 Analysis Uncertainty Factor (γ_{a})

This represents the bias associated with the analytical method (LSP, LDP, or NSP) used to determine the demand measure. It is computed from the following equation:

$$\gamma_{\text{a}} = C_{\text{B}} \exp\left(\frac{k}{2b}\beta_{\text{DU}}^2\right) \tag{5.36}$$

where C_{B} is the bias factor given by the ratio of the demand predicted by nonlinear time-history analysis to the demand predicted by the chosen analytical method. In this case, β_{DU} represents the uncertainty that results from variability in the model parameters such as material strength, assumed damping, hysteretic behavior, and soil–foundation interaction modeling. A series of analyses with different structural models (that account for critical model uncertainties) are carried out using a single ground motion, and statistical measures of dispersion are obtained. β_{DU} is the standard deviation of the natural logarithm of the selected response parameter.

5.6.1.4 Resistance Factor (ϕ)

This factor accounts for the uncertainty in the estimation of the system or component capacity. The resistance factor will have the following general form:

$$\phi = \phi_{\text{R}}\phi_{\text{U}} = \phi_{\text{R}} \exp\left(\frac{k}{2b}\beta^2\right) \tag{5.37}$$

where ϕ_U is the factor accounting for uncertainty in the relationship between laboratory findings and behavior in real buildings (when considering component effects) or the uncertainty in analytical predictions (when considering system response measures such as global drift capacity). FEMA-350 suggests a value of $\beta = 0.2$ for component response measures and a period-dependent value for system response measures. ϕ_R is the randomness inherent in the computation of the resistance factor and is expressed in a form identical to ϕ_U. When computing the resistance factor for component response, β corresponds to the logarithmic standard deviation of the capacity measure based on observed data from laboratory testing. When estimating the resistance factor at the system level, β corresponds to the logarithmic standard deviation of the global stability limit determined from an IDA, as described later.

5.6.2 Simplified Evaluation Method (FEMA-350)

The overall procedure outlined in the Section 5.6.1 is the detailed method that requires numerous simulations of response data and careful statistical analysis of capacity measures. To facilitate routine engineering assessment of regular buildings, FEMA-350 also provides a set of tables to approximately evaluate the uncertainty parameters needed to establish the confidence index parameter. These uncertainty and variability factors are approximate estimates derived from a limited subset of parametric studies based on typical characteristics of regular buildings. Hence, they incorporate a certain degree of conservativeness. Whenever possible, the detailed method described earlier in this section should be used to improve the reliability of the performance evaluation.

A summary of a selected subset of uncertainty factors for special moment frames is shown in Table 5.8 and Table 5.9. Once the uncertainty factors are known, the factored demand-to-capacity index λ can be established for a particular demand measure. Finally, the confidence level of the computed "factored" demand-to-capacity ratio is determined using another table provided in FEMA-350. Table 5.10 shows a slice of the FEMA-350 listing: here, the values tabulated are valid only for midrise frames (4 to 12 stories) at CP performance level. To assess the adequacy of the design, the calculated confidence levels are compared to minimum acceptance criteria. Table 5.11 lists the FEMA criteria for global interstory drift and column compression.

TABLE 5.8 Typical Uncertainty Factors Recommended in FEMA-350 to be Used in Assessing Global Interstory Drift in Special Moment Frames

		Low-rise (<4 stories)	Mid-rise (4–12 stories)	High-rise (>12 stories)
γ_a				
LSP	IO	0.94	1.15	1.12
	CP	0.70	0.97	1.21
NSP	IO	1.13	1.45	1.36
	CP	0.89	0.99	0.95
γ	IO	1.50	1.40	1.40
	CP	1.30	1.20	1.50
ϕ	IO	1.00	1.00	1.00
	CP	0.90	0.85	0.75

TABLE 5.9 Global Interstory Drift Capacity (C) and Associated Uncertainty

		Low-rise (<4 stories)	Mid-rise (4–12 stories)	High-rise (>12 stories)
IO	C	0.02	0.02	0.02
	β_{UT}	0.20	0.20	0.20
CP	C	0.10	0.10	0.085
	β_{UT}	0.30	0.40	0.50

TABLE 5.10 Typical Confidence Levels Listed in FEMA-350 for Computed Factored Demand-to-Capacity Index

Confidence level	10	20	30	40	50	60	70	80	90	95	99
$\beta_{UT} = 0.4$											
λ	2.12	1.79	1.57	1.40	1.27	1.15	1.03	0.90	0.76	0.66	0.51

TABLE 5.11 Sample FEMA-350 Criteria for Assessing Adequacy of Performance

	Performance level	
	Immediate occupancy (IO) (%)	Collapse prevention (CP) (%)
Global behavior limited by interstory drift	50	90
Column compression behavior	50	90

5.6.3 Incremental Dynamic Analysis

The global stability limit corresponds to the CP limit state and is determined using an IDA as proposed by Vamvatsikos and Cornell (2002). An IDA curve is developed from the following procedure:

1. A suite of accelerograms (a minimum of ten records is recommended) is selected representative of the site and hazard level for which the CP level (or global stability limit) is desired. Artificially synthesized records may be used, but it is preferable to choose recorded ground motions that represent the expected hazard at the site in terms of source mechanism, fault distance, and geological characteristics.
2. For each earthquake record, conduct a series of time-history analyses by uniformly scaling the accelerogram each time so as to induce increasing levels of inelasticity in the structure and thereby characterize the behavior of the system from its elastic limit to its failure state.
3. Plot the magnitude of the spectral acceleration of the 5% damped response spectrum of the earthquake record at the fundamental period $S_a(T_0)$ versus the control node (typically the roof) displacement (Δ). Note that the spectral values will scale linearly during the scaling process. The resulting plot is called the IDA curve. Repeat the process for each of the selected ground motions thereby constructing a set of IDA curves.
4. The scale factors used to generate the IDA curve must be carefully selected to capture the elastic, postyield, and collapse limit states. The number of analyses required to develop an IDA curve will vary from one structure to the next; however, Vamvatsikos and Cornell (2002) suggest that the global drift capacity be defined at a point where the slope of the S_a–Δ curve drops below 20% of the elastic slope. Additionally, FEMA-350 places a limit of 10% on the global drift capacity.
5. Establish the distribution of spectral magnitudes at the global drift capacity to facilitate the computation of the logarithmic standard deviation of the drift capacity, β. The global resistance factor is then determined using

$$\phi_R = \exp\left(\frac{k}{2b}\beta^2\right) \tag{5.38}$$

where k and b are the same parameters described in previous expressions.

5.6.4 The PEER PBD Methodology

The methodology in development at the PEER center may be regarded as a holistic approach to PBD in that it involves a measure of performance that is relevant to stakeholders.

A probabilistic framework is used to assess the performance of the structure. This framework is based on a similar format used in the SAC Steel Frame project (Cornell et al. 2002) described in the previous section. While the original formulation by Cornell and coworkers was cast in demand versus capacity format and expressed the performance objective function as the probability of exceeding a certain performance level, the methodology developed for PEER extends this concept to include decision variables and damage measures. A conceptual description of the PEER methodology, using the total probability theorem, is expressed as follows:

$$\nu(\mathrm{DV}) = \iint G(\mathrm{DV}|\mathrm{DM}) \, dG(\mathrm{DM}|\mathrm{EDP}) \, dG(\mathrm{EDP}|\mathrm{IM}) \, d\lambda(\mathrm{IM}) \qquad (5.39)$$

where $\nu(\mathrm{DV})$ is the probabilistic description of the decision variable (expressed in terms of the mean annual probability of repair or replacement cost, net dollar loss, downtime, etc.), DM represents the damage measure (an index value that expresses the state of damage resulting from a given demand), EDP represents the engineering demand parameter (drift, plastic rotation, etc.), and IM represents the intensity measure (characterizing the hazard). The expression of the form $P(A|B)$ is essentially a cumulative distribution function or the conditional probability that A exceeds a specified limit for a given value of B. The term of the form $dP(A|B)$ is the derivative with respect to A of the conditional probability $(A|B)$.

One of the objectives of the PEER methodology is to incorporate all significant sources of uncertainty that arise in the specification of the ground motion, the material properties, and the modeling and evaluation process. The methodology also insists on a rigorous evaluation utilizing state-of-the-art tools for both the prediction of demand and the probabilistic assessment of each random variable. Hence, it would generally be expected that the evaluation be based on nonlinear time-history analyses using a suite of ground motions since it offers more options for the treatment of uncertainties. However, it does not preclude any particular method of analysis as long as the inherent bias and system uncertainties are quantified and probabilistic descriptions of the demand measures are established.

A brief description is now provided of the variables that appear in Equation 5.39.

5.6.4.1 Hazard Description (IM)

This refers to the hazard curve that expresses the annual frequency or exceedance probability of the design seismic event in a specified duration. In developing the hazard model, a rigorous seismic hazard analysis must be carried out that takes into consideration the seismic source, recurrence, and attenuation relations. Once a hazard model is developed, a set of ground motions must be synthesized (artificially) or selected from available records that satisfy the magnitude–distance combinations and geologic and fault characteristics at the site. An intensity measure must then be selected that quantifies the hazard. The idea of introducing an intensity measure to characterize the hazard provides a framework to develop fragility functions. A simple choice of intensity measure would be the PGA. Another option would be the 5% damped spectral acceleration at the characteristic period of the structure ($S_a(T_0)$) since it introduces an additional parameter in the description. There could potentially be numerous such measures with increasing complexity. The goal in selecting a measure of earthquake intensity is to minimize the variability in the performance assessment arising from ground motion parameters.

5.6.4.2 Engineering Demand Parameter (EDP)

The nonlinear analysis of a mathematical model of the structural system produces a wealth of information. The process of selecting one or more demand measures to characterize the performance of the system is not an easy task. The PEER methodology does not identify any specific demand measure since the choice of an appropriate demand parameter is a function of both the system and the desired performance objectives. A performance objective, for example, that attempts to minimize losses in a moderate earthquake might depend on the damage to nonstructural elements. In this case, the demand measure of interest may be interstory drift or peak floor acceleration. On the other hand, if the performance objective were linked to global measures such as LS or CP, it would be prudent to select demand measures that provide a more rational evaluation of the damage states of components. The treatment of modeling uncertainties is important in the process of establishing demands.

5.6.4.3 Damage Measure (DM)

Relevant measures of damage are identified for both structural and nonstructural components. However, unlike deterministic approaches where performance limits are specified for different damage states, the damage description in a probabilistic framework takes the form of a fragility function. The likelihood of a certain damage state is conditioned on a structural response measure or EDP. Similar to the development of acceptance criteria in FEMA-356, a damage model in the PEER framework is expected to rely on damage data from field observations and laboratory testing. Since the database of experimental information is limited, analytical simulations wherein calibrated models can be utilized to generate fragility curves for different variables can aid the development of damage fragilities. Again, this approach provides a means to incorporate material and model uncertainties into the evaluation process.

5.6.4.4 Decision Variable (DV)

The final phase of the PEER methodology involves loss modeling. The terminology that is introduced in this context is a decision variable that is expected to measure system performance in terms of cost estimates, such as mean annual dollar loss (repair costs), or the probability that the damage resulting from the seismic event will result in a specified downtime (loss of operability) and similar economic measures.

The PEER methodology offers distinct advantages over existing PBD frameworks. First, it recognizes the need to incorporate uncertainties in every phase of the process. This enables an assessment of the sensitivity of critical decision variables to uncertain input parameters. Approaches to evaluating system uncertainties and reducing the number of variable parameters that need to be considered have been investigated by Haukaas and Der Kiureghian (2002) and Porter (2002). Second, it provides an explicit probabilistic expression of the system performance in terms of economic measures that are more intelligible to stakeholders. Finally, the methodology is a generic representation of input and output that permits maximum flexibility to the user to define model parameters.

5.7 PBD Application: An Illustrative Example

The basic procedures outlined in this chapter are now applied to a sample six-story steel building to illustrate the application of performance-based concepts in routine seismic evaluation and design. The intention here is to merely demonstrate the process rather than establish a valid design that conforms to a specific performance-based format. As reiterated in several instances in this chapter, PBD is an evolving methodology that still lacks final details on acceptance criteria and loss estimates associated with different damage states. In fact, this example will also serve to illustrate potential unresolved issues in the current state-of-the-art.

For the purpose of this illustration, it will be assumed that the site of the building is located in an active seismic zone in Southern California. The building has a regular and symmetric plan 120 ft^2. The primary lateral load resisting system is a moment frame around the perimeter of the building. Moment continuity of each of the perimeter frames is interrupted at the ends where a simple shear connection is used to connect to the weak column axis. The plan view of the building and the elevation of a typical perimeter frame are shown in Figure 5.14. The interior frames were designed as gravity frames and consist of simple shear connections only. It will be further assumed that the columns of the perimeter frames are supported on foundation beams and rigid pile groups that permit the structure to be modeled as fully restrained at the supports.

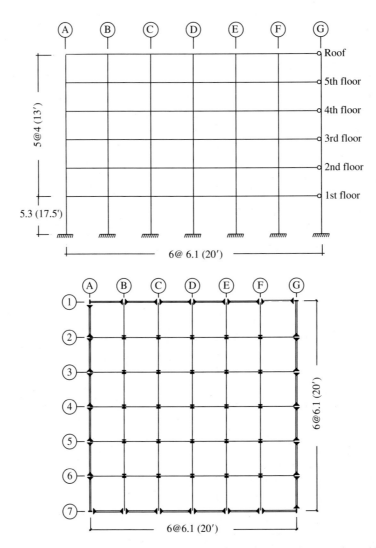

FIGURE 5.14   Building plan and elevation of typical moment frame considered in sample problem.

5.7.1 Performance Objective

The building will be designed for an event with a 10% probability of being exceeded in 50 years. This is referred to as the DE in ATC-40 or a BSE-1 hazard level in FEMA-356. In the present example, a performance level corresponding to LS for all primary structural elements is selected. Secondary or nonstructural performance is not considered in this example. Hence, the performance objective for the trial design is to achieve LS performance in all components when the building is subjected to a DE with the specified probability of occurrence (10% in 50 years).

It is now necessary to define the design event. Since both static and dynamic methods will be used to estimate demands, it is necessary to define a response spectrum and to develop a set of acceleration time histories. To construct the hazard spectrum, the soil profile, the geological characteristics, and source mechanism at the site need to be established. The following values will be assumed to generate the spectrum: seismic zone 4, firm soil with a shear wave velocity exceeding 1000 m/s and the closest distance to a seismic source capable of producing large-magnitude events to be 10 km. The spectral acceleration at a 1.0-s period for site class B is taken as 0.48g. Note that both the short-period and the long-period spectral acceleration values can be determined from USGS seismic maps. This results in the following parameters:

ATC-40:
$$C_A = 0.40g$$
$$C_V = 0.48g$$
$$T_n = C_V/2.5C_A = 0.48 \text{ s}$$
$$T_m = 0.2T_n = 0.096 \text{ s}$$
$$S_t = C_V = 0.48g$$

FEMA-356:
$$S_{xs} = 1.0g$$
$$S_{x1} = S_{m1} = 0.48g$$
$$S_t = S_{x1} = 0.48g$$

Using the parameters defined above, the hazard spectrum for the building site corresponding to a design event with a 10% chance of being exceeded in 50 years is developed and is shown in Figure 5.15.

5.7.2 PBD and Assessment

The next step in the process is to select preliminary sections for the building elements. This stage of the design is similar to current practice wherein preliminary sizing of members must precede the analysis of the building under the imposed seismic loads. In the present example, we will focus on the performance-based design and evaluation of a typical perimeter frame which is designated to carry half the lateral load in each direction. The initial selection of members is shown in Table 5.12. Section properties are

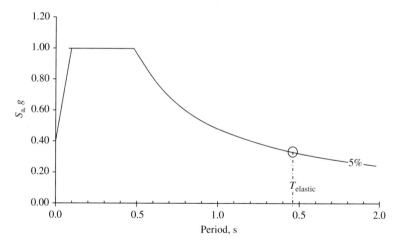

FIGURE 5.15 Design spectrum for building site.

TABLE 5.12 Section Properties of Frame Elements Used in Design Example

Level	Beam section	Column section
6	W24 × 69	W14 × 90
5	W24 × 84	W14 × 90
3–4	W24 × 84	W14 × 132
2	W27 × 102	W14 × 176
1	W30 × 116	W14 × 176

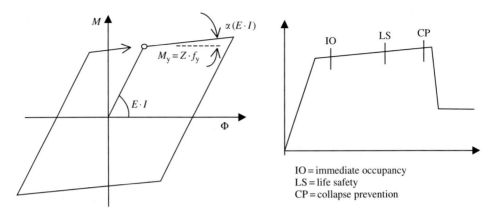

FIGURE 5.16 Prescribed force–deformation behavior of members and monotonic behavior identifying different performance limit states.

computed assuming a yield stress of 414 MPa (60 ksi). Using nominal dimensions for the floor slab and standard weights for nonstructural elements, the floor weights are estimated as 6227 kN (1400 kip) at the first level, 5338 kN (1200 kip) at the second through fifth level, and 7562 kN (1700 kip) at the roof level, resulting in a total building weight of approximately 33,805 kN (7600 kip). A 2D model of the frame was developed to estimate seismic demands. The computer program OpenSees (2003) was used for the analytical simulations since it provided the capability to carry out nonlinear static and dynamic analyses.

Nonlinearity in members is modeled using a bilinear force–deformation behavior in view of the fact that fully restrained connections are used and no brittle fracture is expected. The performance regime of the components will, therefore, lie in the ascending slope of the force–deformation envelope, and softening behavior need not be considered. The expected cyclic behavior of the each yielding section and the monotonic envelope identifying the different performance limit states is qualitatively displayed in Figure 5.16.

Beginning with the trial design (Figure 5.14), the performance of the building is evaluated by determining the demands in each component and comparing these demands to performance-based acceptance criteria. However, as discussed in previous sections of this chapter, there are different approaches to estimating demands and consequently, different assessment criteria to evaluate the performance of the structure. Different methods will be presented in the following sections so as to make readers aware of the predicament that will face engineers when making a decision on which method to use in a given situation. Hopefully, some of the observations and comments offered in this example will assist in clarifying some of these issues.

5.7.2.1 Method I: Deterministic Assessment — LSP (FEMA-356)

The so-called "pseudo" lateral load to be applied on the frame is determined from Equation 5.10. The fundamental period of the building (established from an eigenvalue analysis of the frame model) is 1.45 s. The coefficients needed to evaluate the total lateral load V are

$C_1 = 1.0$ (since $T > T_n$)
$C_2 = 1.0$ for linear procedures
$C_3 = 1.0$ (assuming P–Δ effects can be ignored)
$C_m = 1.0$ (since $T > 1$ s)
$S_a = 0.48/1.45 = 0.33g$

This results in a total lateral force of $V = 6672$ kN (1500 kip) to be applied on each perimeter frame in each direction. The lateral load is distributed over the building height using the expression given in Equation 5.12. The resulting moments on the structural components are recorded, and the peak m-factors are evaluated. Recall that the m-factor is the ratio of the peak moment to the moment capacity (defined here as the yield moment and computed by taking the product of the yield stress of the material and the plastic section modulus) of the beam. A typical summary of such an evaluation is shown in Table 5.13 for all the beam elements on the first level of the building.

This process is carried through all the beams and columns in the structure. The resulting *m*-factors for the entire frame are exhibited in Figure 5.17.

These *m*-factors are then compared to the acceptance criteria listed in FEMA-356 to check if the desired performance level has been satisfied. The limits on *m*-factors for different performance levels are a function of certain section characteristics as identified previously in Table 5.5. For the beams on the first level for which *m*-factor computations were listed in Table 5.13, the following section characteristics are established:

$$b_f = 267 \text{ mm} (10.5 \text{ in.}), \quad t_f = 21.6 \text{ mm} (0.85 \text{ in.}), \quad t_w = 14.5 \text{ mm} (0.57 \text{ in.}), \quad h = 762 \text{ mm} (30 \text{ in.})$$

$$\frac{b_f}{2t_f} = 61.8, \quad \frac{52}{\sqrt{f_{ye}}} = 6.71, \quad \frac{h}{t_w} = 50.6, \quad \frac{418}{\sqrt{f_{ye}}} = 54$$

The above numbers, using Table 5.5, require that for IO performance the *m*-factors be less than 2.0. Since this is true for all beams in the first-story level, they satisfy IO criteria. This performance limit is summarized in Figure 5.18 along with the performance limit states for all elements in the building. As displayed in Figure 5.18, several columns fail LS performance level at the third and fifth floor levels. The notation CP indicates that the element meets the criteria for CP but fails LS performance levels. Since the objective of the design was to ensure LS performance, the result of this evaluation indicates that the

TABLE 5.13 Peak Demands in Beams on First Floor Using LSP

Beam no.	Left	Right	Z	M_y	m-factor Left	m-factor Right	Max, *m*
101	23,670	19,059	378	22,680	1.39	1.12	1.39
102	17,050	17,414			1.00	1.02	1.02
103	17,546	17,507			1.03	1.03	1.03
104	17,437	17,361			1.03	1.02	1.03
105	17,808	19,277			1.05	1.13	1.13
106	11,963	0			0.70	0.00	0.70

							Sixth floor
0.47	0.77	0.76	0.74	0.72	0.64	0.22	
	0.83	0.68	0.66	0.66	0.67	0.46	Fifth floor
0.85	1.26	1.26	1.25	1.23	1.10	0.22	
	1.24	1.02	1.01	1.01	1.06	0.67	Fourth floor
0.71	1.19	1.17	1.17	1.17	1.01	0.08	
	1.51	1.25	1.24	1.23	1.30	0.79	Third floor
0.93	1.39	1.37	1.35	1.34	1.16	0.20	
	1.38	1.11	1.10	1.11	1.19	0.74	Second floor
0.61	1.09	1.37	1.06	1.08	0.94	0.25	
	1.39	1.02	1.03	1.03	1.13	0.70	First floor
1.26	1.39	1.38	1.37	1.36	1.30	0.94	

FIGURE 5.17 Computed *m*-factors from LSP.

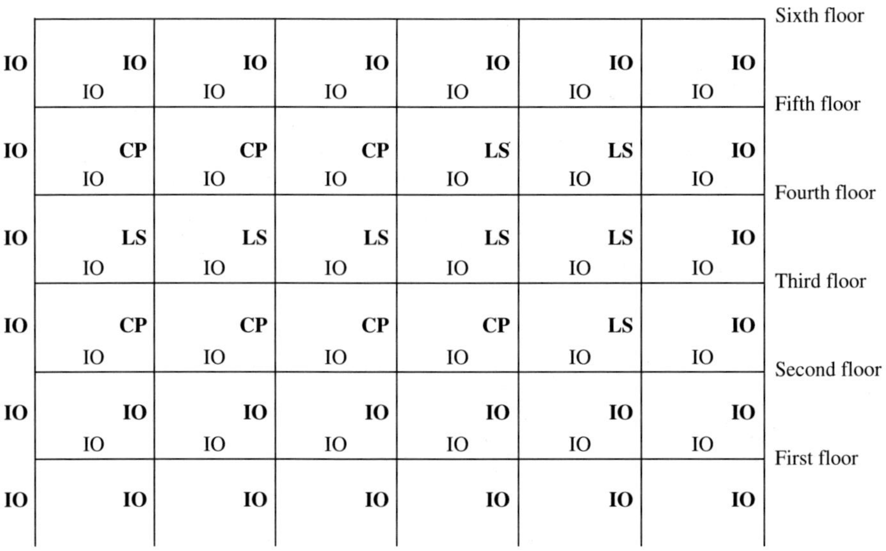

Sixth floor

IO	**IO**	**IO**	**IO**	**IO**	**IO**	**IO**
	IO	IO	IO	IO	IO	IO

Fifth floor

IO	**CP**	**CP**	**CP**	**LS**	**LS**	**IO**
	IO	IO	IO	IO	IO	IO

Fourth floor

IO	**LS**	**LS**	**LS**	**LS**	**LS**	**IO**
	IO	IO	IO	IO	IO	IO

Third floor

IO	**CP**	**CP**	**CP**	**CP**	**LS**	**IO**
	IO	IO	IO	IO	IO	IO

Second floor

IO	**IO**	**IO**	**IO**	**IO**	**IO**	**IO**
	IO	IO	IO	IO	IO	IO

First floor

IO	**IO**	**IO**	**IO**	**IO**	**IO**	**IO**

FIGURE 5.18 Performance assessment of frame based on LSP evaluation.

Note: IO = element passing immediate occupancy criteria; LS = component failing IO but passing life safety criteria; CP = component failing LS but passing collapse prevention criteria.

desired performance objective has not been met. It is, therefore, necessary to redesign the frame and go through an iterative process till the design objective is satisfied.

5.7.2.2 Method II: Deterministic Assessment — NSP (FEMA-356)

The same frame will now be evaluated using a pushover analysis to determine expected demands. The building model is subjected to a monotonically increasing inverted triangular lateral load pattern till the roof drift reaches the target value. The target roof displacement is estimated using the following coefficients:

C_0: 1.42 (From table 3.2 of FEMA-356)
C_1: 1.0 since $T_e > T_S$ (from Hazard Spectra)
C_2: 1.0 (from table 3.3 of FEMA-356)
C_3: 1.0 for positive postyield stiffness

The spectral acceleration at the fundamental period T: 1.45 s is 0.33g. This results in a target displacement of 260 mm (10.22 in.). When the building model is pushed using an inverted triangular lateral load distribution till the roof displacement reaches this value, none of the elements reach their yield value. This means that the plastic rotation demands are zero and, consequently, all elements (beams and columns) satisfy IO performance levels. This is contrary to the findings using linear static analysis.

5.7.2.3 Method III: Deterministic Assessment — LDP (FEMA-356)

A linear dynamic analysis can be accomplished in two ways: using a response spectrum approach or resorting to a full time-history analysis. In this example, the demands were determined using a response spectrum analysis. The moment demands in each element are tabulated, similar to the summary table described for LSP, and the resulting m-factors are calculated. A sample set of values for all the column elements in the first-story level are shown in Table 5.14. For columns, it is also necessary to compute axial force levels since acceptance criteria for columns are a function of the axial demands in the column (see Table 5.14).

TABLE 5.14 Peak Demands in Columns on First Story Using LDP

Column no.	P	A_g	P_{cl}	P/P_{cl}	Bottom	Top	Z	M_y	Co-factor Left	Co-factor Right	Max, m
11	471	56.8	3408	0.14	15,081	10,726	355	21,300	0.94	0.67	0.94
12	157			0.05	16,665	14,123			1.04	0.88	1.04
13	108			0.03	16,524	13,803			1.03	0.86	1.03
14	108			0.03	16,492	13,778			1.03	0.86	1.03
15	119			0.03	16,432	13,749			1.03	0.86	1.03
16	359			0.11	15,777	12,486			0.99	0.78	0.99
17	176			0.05	11,542	3,509			0.72	0.22	0.72

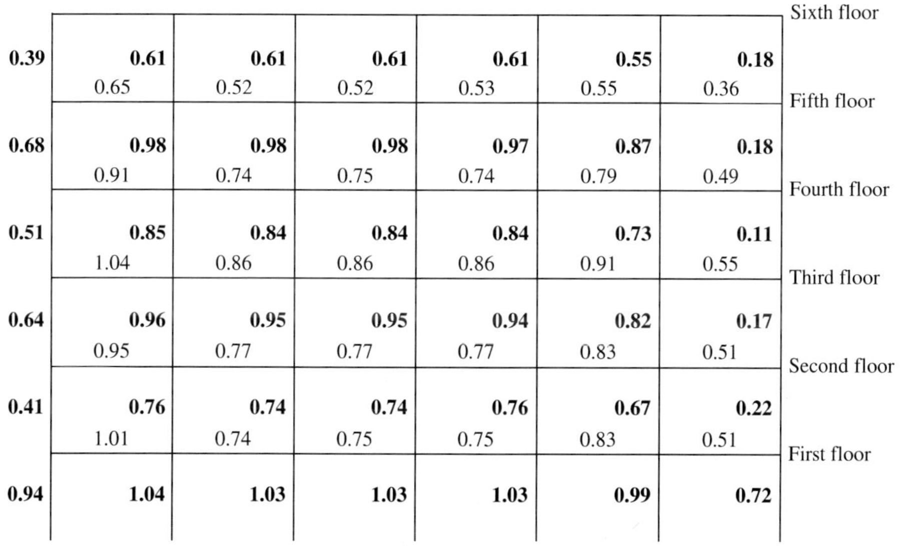

Sixth floor

| **0.39** | **0.61** | **0.61** | **0.61** | **0.61** | **0.55** | **0.18** |
| | 0.65 | 0.52 | 0.52 | 0.53 | 0.55 | 0.36 |

Fifth floor

| **0.68** | **0.98** | **0.98** | **0.98** | **0.97** | **0.87** | **0.18** |
| | 0.91 | 0.74 | 0.75 | 0.74 | 0.79 | 0.49 |

Fourth floor

| **0.51** | **0.85** | **0.84** | **0.84** | **0.84** | **0.73** | **0.11** |
| | 1.04 | 0.86 | 0.86 | 0.86 | 0.91 | 0.55 |

Third floor

| **0.64** | **0.96** | **0.95** | **0.95** | **0.94** | **0.82** | **0.17** |
| | 0.95 | 0.77 | 0.77 | 0.77 | 0.83 | 0.51 |

Second floor

| **0.41** | **0.76** | **0.74** | **0.74** | **0.76** | **0.67** | **0.22** |
| | 1.01 | 0.74 | 0.75 | 0.75 | 0.83 | 0.51 |

First floor

| **0.94** | **1.04** | **1.03** | **1.03** | **1.03** | **0.99** | **0.72** |

FIGURE 5.19 Computed *m*-factors using LDP.

The *m*-factors determined in the table are shown in Figure 5.19 along with the demand-to-capacity values for the remaining elements. Using the criteria listed in FEMA-356, the acceptance levels corresponding to these *m*-factors are established. As can be seen from the *m*-factors, most values are below unity, which means that the demands are generally below the yield capacity of the corresponding components. The demand-to-capacity ratios for a few columns are marginally greater than 1.0. Overall, all elements in the structure pass the criteria for IO. This is consistent with the findings using the NSP.

5.7.2.4 Method IV: Deterministic Assessment — NDP (FEMA-356)

The most rigorous procedure to estimate demands is to carry out a complete nonlinear time-history analysis using a set of ground motions. To accomplish this, a subset of seven earthquake ground motions recommended in ATC-40 is used. Basic seismic parameters of the selected record set are listed in Table 5.15.

As required in FEMA-356, the ground motions are scaled such that the average value of the SRSS spectra does not fall below 1.4 times the 5% damped spectrum for the DE for periods between $0.2T$ and $1.5T$. The spectra for the scaled records along with the original design spectra and the scaled design spectra (wherein the ordinates of the original spectrum are scaled by a factor of 1.4) are shown in Figure 5.20.

TABLE 5.15   Characteristics of Earthquake Records Used in NDP Evaluation

Earthquake no.	Magnitude	Year	Earthquake	Recording station	PGA(g) Horizontal component 1	Horizontal component 2	Vertical component	Distance (km)[a]
1	6.6	1971	San Fernando	Station 241	0.25	0.13	0.17	16.5
2	6.6	1971	San Fernando	Station 458	0.12	0.11	0.11	18.3
3	7.1	1989	Loma Prieta	Hollister, South & Pine	0.18	0.37	0.20	17.2
4	7.1	1989	Loma Prieta	Gilroy #2	0.32	0.35	0.28	4.5
5	7.5	1992	Landers	Yermo	0.15	0.24	0.14	31.0
6	7.5	1992	Landers	Joshua Park	0.28	0.27	0.18	10.0
7	6.7	1994	Northridge	Century City LACC North	0.26	0.22	0.12	23.7

[a] Closest distance to fault.

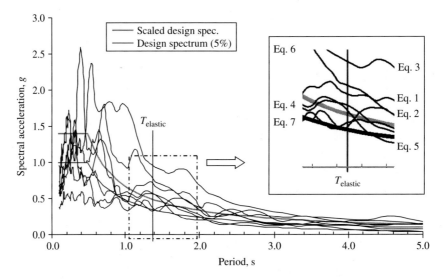

FIGURE 5.20   Original design spectrum, scaled spectrum, and spectra of selected ground motions after scaling.

Component demands from each time-history analysis are recorded, and the maximum values from each earthquake are used to estimate plastic rotation demands. Since seven records were used in the simulation, both FEMA-356 and ATC-40 stipulate that mean values of the peak demands are an adequate measure of the seismic demand. Table 5.16 summarizes the peak and average plastic rotation demands on the columns in the first story for all seven earthquakes. Table 5.16 also shows the maximum demands. This observation is crucial to our understanding of the process governing the choice of earthquake records for an NDP evaluation. If, for example, only three records were used in the evaluation, and Earthquake no. 3 was among the three records, then the seismic demands on the structure would be controlled by the peak values rather than the mean estimates. The maximum demands are considerably different from the mean values, suggesting that engineers should use caution when opting to select mean demands when evaluating performance limits.

In the present case, since all seven data points are used, performance limit states are going to be evaluated on the basis of mean demand estimates. The resulting performance of the building is shown in Figure 5.21. Though only the mean values were used, the NDP results indicate that several columns do not pass IO criteria but do pass LS criteria. Since LS performance is achieved for all elements, the trial design is acceptable.

TABLE 5.16 Peak and Mean Plastic Rotation Demands on First-Story Columns Using NDP

Column no.	Earthquake no.							Mean	Max, *m*
	1	2	3	4	5	6	7		
11	0.14	0.00	1.15	0.00	0.00	1.08	0.00	0.34	1.15
12	0.41	0.00	1.78	0.14	0.23	1.56	0.00	0.59	1.78
13	0.39	0.00	1.71	0.12	0.20	1.59	0.00	0.57	1.71
14	0.38	0.00	1.70	0.11	0.19	1.58	0.00	0.57	1.70
15	0.36	0.00	1.65	0.11	0.18	1.54	0.00	0.55	1.65
16	0.24	0.00	1.74	0.00	0.19	1.30	0.00	0.50	1.74
17	0.00	0.00	0.00	0.00	0.00	0.00	0.00	0.00	0.00

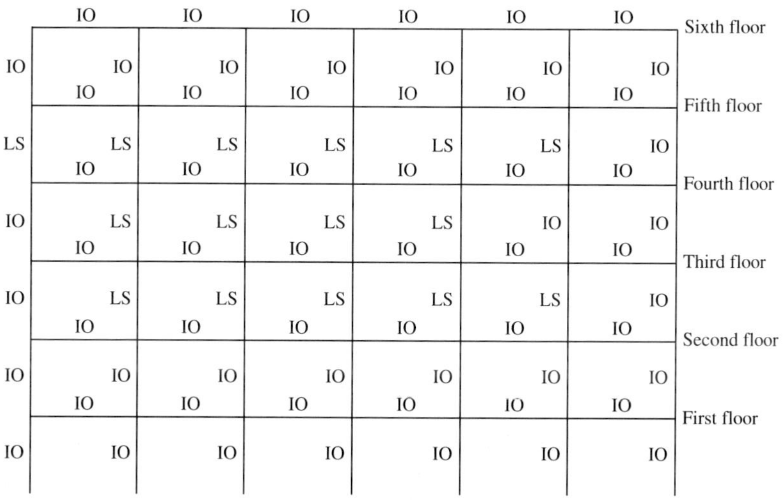

FIGURE 5.21 Performance assessment of frame based on LDP evaluation (see notation in Figure 5.18).

5.7.2.5 Method V: Probabilistic Assessment — LSP (FEMA-350)

The previous four methods provided a deterministic assessment of the building performance. The same design will now be evaluated in a probabilistic manner using the criteria in FEMA-350. The process itself is not probabilistic, though the final evaluation results in a nondeterministic evaluation of the computed demands. We begin with the LSP. The steps involved in predicting demands are similar to method I. The total lateral force is applied as an inverted triangular load using the distribution given by Equation 5.11, and the resulting moment demands in each component are tabulated. These demands are then factored by the uncertainty factors (see Table 5.8) to establish the factored demand value $\gamma\gamma_a D$. Acceptable capacity values are then determined (see typical capacity limits given in Table 5.9) and modified by a capacity reduction factor (shown in Table 5.8), to account for the inherent uncertainty in establishing such limits. The factored demand-to-capacity ratio, λ, is used to establish a confidence level. The computed confidence levels for global interstory drift are tabulated in Table 5.17. The acceptability of the estimated confidence level is specified in FEMA-350 as follows:

Recommended Minimum Confidence Levels (%)

	IO	CP
Global behavior limited by interstory drift	50	90
Local connection behavior limited by interstory drift	50	50
Column compression behavior	50	90
Column splice tension behavior	50	50

TABLE 5.17   Confidence Levels from Factored Demand and Capacity Estimates (LSP)

Story level	IS drift	$(\gamma\gamma_a)D$ IO	$(\gamma\gamma_a)D$ CP	λ IO	λ CP	Confidence level IO	Confidence level CP
0							
1	0.014	0.023	0.027	1.15	0.31	35	99
2	0.014	0.022	0.026	1.11	0.31	47	99
3	0.017	0.028	0.033	1.40	0.38	10	99
4	0.016	0.026	0.030	1.29	0.35	20	99
5	0.015	0.024	0.028	1.19	0.33	30	99
6	0.009	0.014	0.017	0.72	0.20	98	99

TABLE 5.18   Computed Confidence Levels Using NSP

Story level	IS drift	$(\gamma\gamma_a)D$ IO	$(\gamma\gamma_a)D$ CP	λ IO	λ CP	Confidence level IO	Confidence level CP
0							
1	0.037	0.074	0.088	3.71	1.04	< 10	69
2	0.029	0.059	0.070	2.95	0.82	< 10	87
3	0.043	0.087	0.103	4.34	1.21	< 10	55
4	0.035	0.072	0.085	3.58	1.00	< 10	72
5	0.026	0.052	0.062	2.62	0.73	< 10	92
6	0.010	0.021	0.025	1.05	0.29	< 52	99

Based on the recommended acceptance criteria, it is deduced that the confidence level associated with preventing building collapse at the computed drift values is extremely high (actually, this value is at its upper limit). However, the confidence level on IO performance is unacceptable. The minimum recommended value in FEMA-350 is 50%, while the confidence levels based on computed interstory drift are well below these limits for the first five story levels. Unfortunately, FEMA-350 does not specify confidence levels for LS performance. It will be necessary to use the detailed approach to establish the confidence level for this performance state. Consequently, it will not be feasible to evaluate the performance objective of this example using FEMA-350.

5.7.2.6   Method VI: Probabilistic Assessment — NSP (FEMA-350)

Though we have seen that FEMA-350 cannot be used to evaluate LS performance using the approximate method, the evaluation is extended in this section to include NSPs. The intent here is to provide a comparison of the performance evaluation using the same guideline (in this case FEMA-350) but employing different analytical methods to estimate seismic demands. It was shown earlier that the FEMA-356 evaluation using LSP resulted in unacceptable performance, while all the remaining procedures indicated satisfactory performance for the stated design objective.

Results of the evaluation using a pushover analysis (or NSP) are given in Table 5.18. The confidence levels are unacceptable for IO performance at all levels and for CP performance at the third story-level. It is observed that the confidence levels are generally lower than those estimated using LSP. This is interesting because the drift values using NSP are identical for both FEMA-356 and FEMA-350. However, FEMA-356 determines acceptance criteria using local component demands only while FEMA-350 uses a multilevel acceptance criterion. In the present case, global interstory drift was used to evaluate the building performance.

5.7.2.7 Method VII: Capacity Spectrum Based Evaluation (ATC-40)

The next evaluation methodology that will be discussed is the CSM outlined in ATC-40. The first step is to establish the capacity curve, which in this case is simply the pushover curve for the building. However, unlike the process discussed in Method II, no target displacement is used. Instead, the lateral loads are applied incrementally until a stability point is reached (impending $P–\Delta$ collapse). Since a nonlinear computer program is employed for the analysis, the approximate technique (Figure 5.11) discussed earlier does not apply. The pushover curve, which is obtained in terms of base shear and roof displacement, is converted into ADRS format (Equations 5.20 and 5.21). The next step is to select a trial performance point. Before doing this, it is instructive to overlay the 5% damped spectrum, also transformed into ADRS format, with the capacity curve to get a sense of the demand region. Figure 5.22 shows the original demand spectrum and the pushover capacity spectrum. The intersection of the demand and capacity curves occurs at approximately 180 mm (7.1 in.). At this level of displacement, the system is still elastic and the effective period is still 1.45 s. This implies that the intersection point is indeed the performance point since introducing additional damping will only reduce the demand. This result is consistent with the finding using Method II, which also concluded that all members will remain in the elastic range when the structure is subjected to design lateral forces.

5.7.2.8 Method VIII: PEER Probabilistic Framework for Performance Assessment

The final evaluation method to be considered is the PEER probabilistic approach. The first step in the PEER methodology calls for a site-specific analysis to develop the probabilistic hazard curve. As outlined in the detailed FEMA-350 procedure, it will be necessary to determine the slope of the hazard curve to carry out the integration implied in the PEER methodology (Equation 5.39). This means that we need to define at least two hazard levels. In this case, the following hazard levels are selected: events with a 10% probability of being exceeded in 50 years and events with a 2% probability of being exceeded in 50 years. To create a new hazard level beyond that specified in the design objective, the design spectrum for the 10%/50 hazard is scaled by a factor of 1.5 to produce the 2%/50 hazard spectrum. Earlier, a set of seven records corresponding to a 10%/50 hazard level was used in the FEMA-356 NDP evaluation. The same seven records are scaled in a manner consistent with the criteria specified in ATC-40 and FEMA-356 to produce records for the evaluation of the building model for 2%/50 hazard. The design spectral curves and the mean and median of the seven records for both hazard levels are displayed in Figure 5.23.

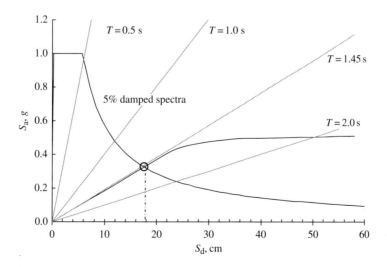

FIGURE 5.22 Capacity spectrum analysis of the building using ATC-40.

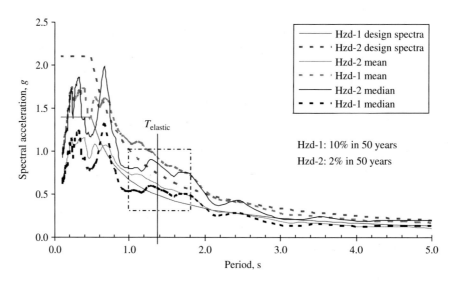

FIGURE 5.23 Mean and median spectra of scaled ground motions for both hazard levels.

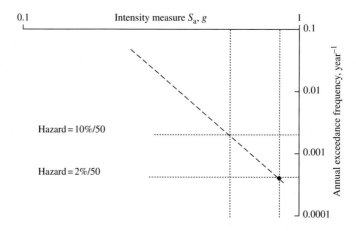

FIGURE 5.24 Probabilistic hazard curve developed for building location.

The intensity measure selected to characterize the hazard of each earthquake is the spectral acceleration at the fundamental period, $S_a(T)$. Ground motions for each hazard level now need to be scaled so as to match the magnitude of the spectral accelerations of the hazard spectrum at the fundamental period. To remain reasonably consistent with the FEMA-356 NDP evaluation, the median curves are used to represent the hazard spectra. Recall that in the NDP evaluation, the ground motions were scaled such that the average value of the SRSS spectra does not fall below 1.4 times the spectral magnitudes of the DE for periods between $0.2T$ and $1.5T$. The resulting hazard curve, plotted in log–log scale, is shown in Figure 5.24. It is evident that an expression of the following form, as suggested in Equation 5.34, can be developed:

$$v(S_a) = k_0(S_a)^{-k}$$

Next, the expected demands in the system, both at the global and at the local level, need to be estimated. The NDP is selected as the analysis method to estimate these demands. The model of the building is subjected to a total of 14 earthquake records, seven for each hazard level. The peak interstory drifts resulting from the nonlinear dynamic analyses are summarized in Figure 5.25.

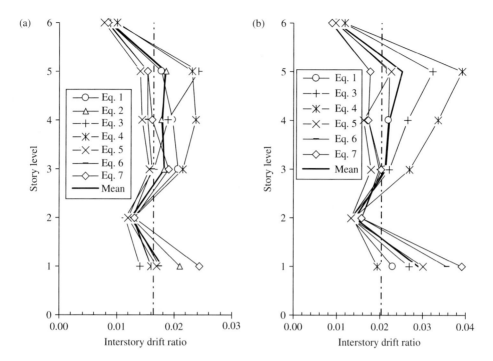

FIGURE 5.25 Interstory drift across building height for both hazard levels.

For the FEMA NDP evaluation, only the peak plastic rotation demands were needed to assess the performance of the system. In the PEER approach, the choice of a demand measure is left to the discretion of the user. For example, we could choose to use peak plastic rotations in any component at a given story level. However, in this case, the peak interstory drift ratios (IDR) are utilized to evaluate performance. In general, it would be advisable to evaluate system performance using several demand measures ranging from global measures such as drift to local measures such as plastic rotation demand. The distribution of the IDR as a function of the intensity measure is shown in Figure 5.26. Only the peak IDRs for each analysis are used to generate this plot. It is now possible to develop a drift hazard curve of the following form:

$$e^{\mu_{\ln[x|S_a]}} = a\left(S_a\right)^b \tag{5.40}$$

where $\mu_{\ln[x|S_a]}$ is the mean value of the natural logarithm of the IDRs associated with the given intensity measure and a and b are constants to be evaluated through curve-fitting the data.

The objective in this phase of the evaluation is to estimate the probability of occurrence of a peak IDR conditioned on a given spectral ordinate, as follows:

$$P(x|\text{IM} = S_a) = \Phi\left(\frac{\ln(x) - \mu_{\ln[x]}}{\sigma_{\ln[x|S_a]}}\right) \tag{5.41}$$

where x is the peak IDR, IM is the intensity measure which has been selected as $S_a(T)$, $\mu_{\ln[x]}$ is as defined previously, and $\sigma_{\ln[x|S_a]}$ is the standard deviation of the natural logarithm of the IDRs at the same intensity measure. If the drift demand is considered to be a random variable, and the drift hazard curve is established as described in Equation 5.40, Luco and Cornell (1998) show that the annual exceedance probability can be estimated from

$$v(x) = k_0 \left[\left(\frac{x}{a}\right)^{1/b}\right]^{-k} \exp\left[0.5k^2 \left(\frac{\sigma_{\ln(x|S_a)}}{b}\right)^2\right] \tag{5.42}$$

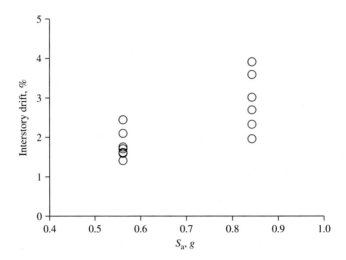

FIGURE 5.26 Distribution of peak interstory drift as a function of earthquake intensity measure.

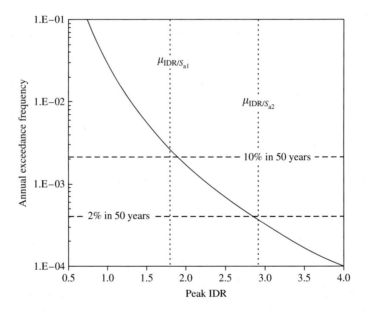

FIGURE 5.27 Interstory drift hazard.

All terms in the above expression have been previously defined. Since we have two distributions of the IDRs for two hazard levels, separate standard deviations are available for each hazard. However, for the design example under consideration, the dispersion associated with the 10%/50 year hazard is utilized to construct the plots of exceedance probability shown in Figure 5.27.

The information presented in Figure 5.27 is only an intermediate step in the PEER methodology. Next, we need to establish a damage definition given a demand value. There are obviously many ways to do this. For the purpose of this study, we will examine the use of IDA to identify critical limit states and use these limits to define a measure of damage. For example, IDA curves are recommended in FEMA-350 to determine the collapse limit state of the building. In this example, the IDA curve is used to specify several possible limit states. In the IDA analysis, each of the seven earthquake records is scaled repeatedly to produce a range of demand values from the elastic state to near collapse. The collapse limit state in most applications is a stability limit state. When using a nonlinear computer program with ability to incorporate *P*–Δ effects, this limit state corresponds to numerical instability (or inability to converge within the specified tolerance) arising from *P*–Δ softening.

Results of the IDA evaluation are displayed in Figure 5.28. For each analysis, both the roof drift and the peak interstory drift values are recorded. In many cases, failure may be triggered by local soft stories, and damage states are more discernable when examining story drift measures rather than global response measures. An interesting feature of IDA is that it provides a sense of the complexity of inelastic dynamic response. Scaling the accelerogram does not necessarily result in increased demands. Each earthquake produces a different pattern of demands. Hence, it is necessary to consider a statistically significant sample of earthquake records to extract meaningful information from an IDA.

Since the primary objective of this example is to demonstrate the methodology rather than offer a complete evaluation of the building, the IDA curves presented in Figure 5.28 will now be evaluated to define performance limit states. The response of the system is essentially in the elastic range up to about 1% peak interstory drift (at any level). However, damage to the contents in the building may occur at lower drift limits. A deviation from the initial linear portion of the curve is evident around 2% for most records. Earthquake No. 2 produces considerable softening in the system beyond 2% interstory drift. Based on these qualitative observations, the following limit states are defined:

- *Immediate occupancy limit state.* IDR of 1.0%.
- *Life safety limit state.* IDR of 3.0%.

Assuming that the above IDR values represent the mean estimates and using the dispersion (standard deviation) in the IDR data at the above two limit states, the damage probabilities conditioned on the IDRs are determined and displayed in Figure 5.29a. It is important to reiterate that the above limit states are rather arbitrary and correspond to global response measures only. Detailed evaluation of demands at the component (and nonstructural component) level is necessary to develop more rational limit states in an actual building evaluation. For the present example, therefore, if the above limit states are defined, it is possible to assess the building performance from a probabilistic standpoint. If the PEER methodology is carried through as suggested in Equation 5.33, it is also necessary to define a decision variable. A careful examination of the different variables that appear in the

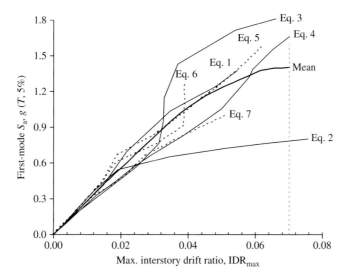

FIGURE 5.28   IDA curves for the six-story frame using the records listed in Table 5.15 (peak IDR at stability limit is identified as 0.07).

PEER equation indicates that it may not be necessary to integrate across three separate variables. For example, a demand measure may be translatable directly into a decision variable since damage parameters and decision variables are closely interlinked. In this illustration, the evaluation will not be carried through to the level of a decision variable (meaning a quantified economic loss measure) as intended in the PEER methodology primarily because the task of defining and calibrating such variables is beyond the scope of this chapter. Instead, the damage fragility will be established by considering the individual probabilities of damage and demand in the following manner:

$$P(\mathrm{DM}|S_a) = \int P(\mathrm{DM}|x)P(x\,|\,S_a)\ \mathrm{d}x \tag{5.43}$$

The variable x in the above expression is the IDR. $P(\mathrm{DM}|x)$ is plotted in Figure 5.29a. Combining this with $P(x\,|\,\mathrm{IM})$ and carrying out the assessment indicated by Equation 5.43 results in the fragility curves shown in Figure 5.29b. The results indicate that the likelihood of exceeding 1% drift (which we have defined to correspond to IO performance criteria) is almost certain if the spectral acceleration of the ground motion (based on constructing a 5% damped elastic spectrum of the accelerogram) exceeds about 0.6g at a fundamental period of approximately 1.39 s. The required magnitude of the spectral acceleration at the fundamental period of the building for LS performance is about twice as much.

5.7.3 Notes and Observations

The primary purpose of the illustrative exercises presented previous subsections was to highlight issues in the current state-of-the-art in performance-based seismic design. There are clearly many unresolved issues, particularly pertaining to acceptance criteria and method of analysis. For example, in the sample evaluations discussed here, the findings were inconsistent for different methods. Using the same design spectrum, the demands from different methods produced different performance levels. With reference to FEMA-356, it is observed that in this case the linear static method was the most conservative. The remaining three methods indicated satisfactory behavior with the performance objective being achieved. However, the results from LDP and NSP indicated an essentially elastic response, while demands from NDP showed inelastic behavior in some elements though the demands were within LS levels. Analysis using ATC-40 and CSMs was similar to FEMA-356 NSP. In the case of FEMA-350, which is written in a format conforming to modern code specifications, no criteria are specified for LS performance.

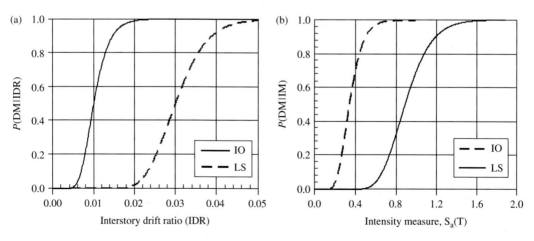

FIGURE 5.29 Fragility functions showing damage probabilities for two performance states: (a) damage probability conditioned on IDR and (b) damage probability conditioned on intensity measure.
Note: IO and LS limit states correspond to 1% and 3% IDR, respectively.

However, an interesting observation here is that an evaluation using NSP was less conservative than that with LSP, which is contrary to a similar comparison using FEMA-356. The results of the evaluation using the PEER methodology cannot be directly compared to the results of the previous procedures. Yet, it is possible to arrive at a nondeterministic understanding of the response: if we were to use the spectral acceleration of 0.58g corresponding to the 10%/50 year hazard, then the probability of achieving IO performance level is low (in other words, there is a high probability that the 0.01 IDR threshold for IO performance will be exceeded). This is based on the conclusion from the IDA curve that suggests a 0.01 peak IDR to correlate with IO performance level (or likely elastic limit). However, the likelihood of more severe damage states decreases dramatically (for example, Life Safety performance at the 10%/50 year hazard level is easily achieved since the probability of exceeding 3% drift corresponding to LS performance is nearly zero). Ideally, it should also be possible to develop loss estimates associated with different damage states. The reader is referred to the work of Miranda and Aslani (2003) and Krawinkler and Miranda (2004) for a more in-depth treatment of this subject.

5.8   Concluding Remarks

While much progress has been made in the last decade to develop processes and procedures that incorporate performance-based concepts in design and evaluation of structural systems, the findings and observations presented in the previous section raise important issues that point to the need for additional research and the development and validation of methodologies that minimize uncertainty and maximize confidence. The ongoing work at PEER is clearly a major advancement that deserves attention. Another ongoing effort is the FEMA-sponsored ATC-58 project, which has been charged with developing the next-generation performance-based seismic design procedures and guidelines. The objective of the ATC endeavor is to "express performance directly in terms of the quantified risks" in a format that is comprehensible to the decision maker. The project is expected to be conducted in two phases: in the first phase a set of procedures will be developed using probabilistic methods to assist an engineer to evaluate either an existing building or the proposed design of a new building utilizing terminology defined by decision makers; in the second phase a complimentary set of guidelines for use by stakeholders will be developed.

Since PBEE seeks to define a range of damage states, considerable work remains to be done to develop advanced computational tools that predict seismic demands of a complete soil–foundation–structure system reliably. Precise modeling guidelines should accompany such tools. Work is needed to quantify damage states and to develop models that transform damage states into loss estimates. Methods of analysis need to be validated with more rigor, particularly simplified methods that can be used in lieu of fully nonlinear time-history evaluations. Pushover procedures, which work well for low-rise, regular structures must be enhanced so that they can be used for a wider range of structural configurations. The work of Gupta and Kunnath (2000) and Chopra and Goel (2002) are important developments in this area. But more importantly, the propagation of uncertainties must be incorporated through the demand and assessment process. The degree of sophistication or level of detail will depend on the application, on whether the structure is classified as being ordinary or critical, and numerous other factors.

The success of a new seismic design methodology lies in the success of its implementation in building practice. Based on patterns in other earthquake innovations (e.g., base isolation and load and resistance factor design (LRFD)) May (2000) suggests that it may take at least two decades for ideas to move from the initial onset of preliminary guidelines (such as FEMA-356 and FEMA-350) to widespread adoption. Some of the key obstacles to advancing and implementing performance-based earthquake engineering (PBEE) cited by May include uncertainties in the methodology and its benefits, costs associated with adopting the methodology, complexity of the methodology in comparison to current code format, and issues related to validating and facilitating the adoption of the methodology. Another factor that can impact the development of PBEE is the interest of stakeholders in the success of the methodology. Steps to mitigate these barriers are essential to hasten the adoption of PBEE.

In conclusion, it must be reiterated that the intent of this chapter is to provide readers with a glimpse of evolving views on PBSE so that they can be better prepared as the first performance-based code for seismic design becomes a reality in the future. At a time when performance-based concepts were still being debated, Krawinkler (1997) noted that PBEE "appears to promise engineered structures whose performance can be quantified and conform to the owner's desires. If rigorously held to this promise, performance-based engineering will be a losing cause." This prediction can be regarded as ominous or enlightening depending on the perspective of the reader. Rather than offer an unpromising indictment, Krawinkler's purpose was to underscore the myriad of uncertainties associated with the overall process. In fact, he continues by saying that "significant improvements beyond the status quo will not be achieved without a new and idealistic target to shoot for. We need to set this target high and strive to come close to its accomplishment . . . Performance based seismic engineering is the best target available and we need to focus on it." The real challenge in PBSE is not simply to better predict seismic demands or improve our loss-assessment methodology, but it is "in contributing effectively to the reduction of losses and the improvement of safety." (Cornell and Krawinkler 2000).

Appendix A

Conceptual Basis of Pushover Analysis and CSM

In this section, a few conceptual details on some of the processes and methods discussed in this chapter are outlined. These concepts should aid the reader in comprehending the basis of both pushover analysis and CSMs. Figure 5.A1 shows an idealized model of a four-story frame. In a standard computer analysis of the 2D frame model, each node will be assigned three DOFs corresponding to axial, shear, and bending deformations. However, the primary degree of interest in pushover analysis and CSMs is the lateral floor DOF since it is generally assumed that floors are rigid and a SDOF is adequate to represent the drift at a story level. When constructing the system stiffness matrix, all three DOFs are used; however, only the floor DOFs are retained (the remaining DOFs are condensed, not eliminated or ignored).

The idea behind static pushover methods is to characterize the nature of seismic loads acting on the frame. The response of the frame to any dynamic load is a combination of the dynamic modes of vibration of the system. Hence, it is instructive to begin with fundamental concepts in modal analysis of structures. As is well known, any vector of order n can be expressed by a set of n independent vectors. In this case, the independent eigenvectors resulting from the solution of the eigenvalue problem serve as appropriate vectors to express the floor displacements of a multistory building. The variable n refers to the number of DOFs which in this case is the number of floor levels for the reasons cited:

$$\{u_i\}_m = \sum_{m=1}^{N} \Phi_m q_m = [\Phi]\{q\} \tag{5.A1}$$

where $\{u_i\}$ is the displacement vector, $\{q\}$ is the normal or modal coordinates, $[\phi]$ is the matrix of eigen vectors, and m and i are the mode number and floor level, respectively.

Let us now consider the equilibrium relationship for a multi-DOF system:

$$[m]\{\ddot{u}\} + [c]\{\dot{u}\} + [k]\{u\} = -[m]\{\imath\}\ddot{u}_g(t) \tag{5.A2}$$

where $[m]$, $[c]$, and $[k]$ are the mass, damping, and stiffness matrices, respectively; $\{u\}$, $\{\dot{u}\}$, and $\{\ddot{u}\}$ are the displacement, velocity, and acceleration vectors, respectively; $\{\imath\}$ is a vector of unit values; and $\ddot{u}_g(t)$ is the imposed ground motion acceleration.

Utilizing the modal decomposition given in Equation 5.A1 and applying orthogonality relationships, the equilibrium expression reduces to the following form:

$$\ddot{q}_n + 2\zeta\omega_n\dot{q}_n + \omega_n^2 = -\Gamma_n\ddot{u}_g(t) \tag{5.A3}$$

where $\Gamma_n = ([\Phi]^T[m]\{\imath\})/M_n$, in which $M_n = [\Phi]^T[m][\Phi]$.

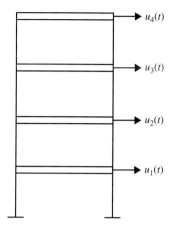

FIGURE 5.A1 Typical multistory frame model with reduced DOFs.

Actually, a more informative way of expressing the right-hand side of Equation 5.A2 is to consider independent modal contributions as presented by Chopra (2001):

$$[m]\{\ddot{u}\} + [c]\{\dot{u}\} + [k]\{u\} = -\sum_{n=1}^{N} R_n \ddot{u}_g \tag{5.A4}$$

Comparing Equation 5.A4 to Equation 5.A2 and following through with the modal transformation that results in Equation 5.A3, it can be shown that

$$\{R\} = \sum R_n = \Gamma_n m \Phi_n \tag{5.A5}$$

Each term in the above expansion contains the modal contribution of the respective mode. Another way of visualizing Equation 5.A5 is to consider the load vector on the right-hand side of Equation 5.A2 as follows:

$$[m]\{\iota\}\ddot{u}_g = \{R\}f(t) \tag{5.A6}$$

where $\{R\}$ is a load distribution vector. For a general loading function $\{p(t)\} = \{r\}f(t)$, the vector $\{r\}$ represents a displacement transformation vector resulting from a unit support displacement. For earthquake loading, this simply becomes a vector with unit values. The external loading can obviously vary as a function of time in terms of both amplitude and spatial distribution. The objective of deriving an expression of the form given by Equation 5.A6 is to separate the spatial distribution from the time-varying amplitude function. This concept is not new and has been discussed in standard textbooks in dynamics (Clough and Penzien 1993).

The next step is to introduce features of the earthquake loading. Since the procedure being developed is a static procedure, the most appropriate form of earthquake loading that can be considered is a response spectrum. The spatial distribution of lateral forces to be used in conjunction with a pushover analysis is approximated in terms of the peak modal contributions as follows:

$$\{f_n\} = \Gamma_n[m]\{\Phi_n\}S_a(\zeta_n, T_n) \tag{5.A7}$$

where S_a is the spectral acceleration for the given earthquake loading at a frequency corresponding to the period T and damping ratio ζ for mode n.

The modal forces computed using Equation 5.A7 will represent the contributions to mode n only. Equation 5.A7 represents the most general form of the lateral force vector to be used in a pushover analysis. If $n = 1$, only the first mode contributions are considered. In the approach proposed by Chopra and Goel (2002), it is assumed that the inelastic response can also be approximated by modal super-position because the *n*th mode is expected to be dominant even for inelastic systems. Other modal combination techniques (Kunnath 2004) and even adaptive lateral force distributions have been proposed (Gupta and Kunnath 2000) based on interesting variations in applying the load represented by Equation 5.A7. These concepts are beyond the scope of this chapter.

To understand the concept of a capacity spectrum and the ADRS conversion, it is necessary to return to Equations 5.A1 to 5.A3 and follow through with the general modal analysis procedure. The peak response of an SDOF system subjected to a given ground motion can be obtained from a response spectrum of the ground motion. Equation 5.A3 describes a set of n SDOF systems with each expression providing the solution to a specific mode. The total response is obtained through the transformation given by Equation 5.A1.

Assuming $S_d(\zeta_n, \omega_n)$ to represent the maximum displacement of an SDOF system with frequency ω_n and damping ratio ζ_n when subjected to the ground motion $\ddot{u}_g(t)$, the peak displacement response of the system represented in Equation 5.A3 is given by

$$\{q_n\}_{\max} = \Gamma_n S_d(\zeta_n, \omega_n) \tag{5.A8}$$

The peak story displacements can now be determined using Equation 5.A1 as follows:

$$\begin{Bmatrix} u_1 \\ u_2 \\ \vdots \\ u_n \end{Bmatrix}_{\max} = \Gamma_1 S_d(\zeta_1, \omega_1) \begin{Bmatrix} \Phi_{11} \\ \Phi_{21} \\ \vdots \\ \Phi_{n1} \end{Bmatrix} + \Gamma_2 S_d(\zeta_2, \omega_2) \begin{Bmatrix} \Phi_{12} \\ \Phi_{22} \\ \vdots \\ \Phi_{n2} \end{Bmatrix} + \cdots + \Gamma_n S_d(\zeta_n, \omega_n) \begin{Bmatrix} \Phi_{1n} \\ \Phi_{2n} \\ \vdots \\ \Phi_{nn} \end{Bmatrix} \tag{5.A9}$$

The above expression contains the contributions of all modes. Let us assume that only the peak displacement at a particular DOF is needed. For example, if DOF n represents the roof level, and only the first-mode contribution is considered, then the following expression is obtained:

$$u_{n,\max} = \Gamma_1 S_d(\zeta_1, \omega_1) \Phi_{n1} \tag{5.A10}$$

This equation is used to convert the roof displacement resulting from a pushover analysis to the first-mode spectral displacement in the capacity spectrum procedure.

To establish the equivalent first-mode spectral acceleration from the base shear, the peak displacement

$$\{f_n\}_{\max} = \omega_n^2 [m]\{u_n\}_{\max} \tag{5.A11}$$

$$\{f_n\}_{\max} = \omega_n^2 [m]\Gamma_n S_d(\zeta_n, \omega_n)[\Phi] \tag{5.A12}$$

$$\{f_n\}_{\max} = \Gamma_n S_a(\zeta_n, \omega_n)[m][\Phi]$$
$$= \left\langle \Gamma_1 S_a(\zeta_1, \omega_1) \begin{Bmatrix} m_1\Phi_{11} \\ m_2\Phi_{21} \\ \cdots \end{Bmatrix} + \Gamma_2 S_a(\zeta_2, \omega_2) \begin{Bmatrix} m_1\Phi_{12} \\ m_2\Phi_{22} \\ \cdots \end{Bmatrix} + \cdots \right\rangle \tag{5.A13}$$

If only the first-mode contribution is considered

$$\{f_n\}_{\max} = \Gamma_1 S_a(\zeta_1, \omega_1) \begin{Bmatrix} m_1\Phi_{11} \\ m_2\Phi_{21} \\ \cdots \end{Bmatrix} \tag{5.A14}$$

The base shear is the sum of story forces, hence the first-mode contribution to the base shear is given by

$$V = \Gamma_1 S_a(\zeta_1, \omega_1) \sum_{i=1}^{n} m_i \Phi_{i1} \tag{5.A15}$$

Acknowledgments

I would like to acknowledge the assistance and contributions of Erol Kalkan in preparing the design example presented in this chapter. Conversations with Helmut Krawinkler on performance-based engineering and with Eduardo Miranda on the PEER methodology, provided new perspectives that have aided my thought process in the preparation of this chapter.

References

ATC 3-06 (1978). *Tentative Provisions for the Development of Seismic Regulations for Buildings.* Applied Technology Council, Redwood City, CA.

ATC-40 (1996). *Seismic Evaluation and Retrofit of Concrete Buildings.* Report SSC 96-01, California Seismic Safety Commission, Applied Technology Council, Redwood City, CA.

Cheok, G., Stone, W., and Kunnath, S.K. (1998). Seismic response of precast concrete frames with hybrid connections. *ACI Struct. J.* 95(5), 527–539.

Chopra, A. (2001). *Dynamics of Structures: Theory and Applications to Earthquake Engineering.* Prentice Hall, New York.

Chopra, A. and Goel, R. (2001). Direct displacement-based design: use of inelastic vs. elastic design spectra. *Earthquake Spectra* 17(1), 47–64.

Chopra, A. and Goel, R. (2002). A modal pushover analysis procedure for estimating seismic demands for buildings. *Earthquake Engineering and Structural Dynamics* 31 (3), 561–582.

Clough, R. and Penzien, J. (1993). *Dynamics of Structures.* McGraw-Hill, New York.

Cornell, C.A. (1996). Calculating building seismic performance reliability: a basis for multilevel design norms. *11th World Conference on Earthquake Engineering*, Paper No. 2122. Elsevier Science Ltd., Amsterdam.

Cornell, C.A. and Krawinkler, H. (2000). Progress and challenges in seismic performance assessment. *PEER Center News* 3(2), 1.

Cornell, C.A., Jalayer, F., Hamburger, R.O., and Foutch, D.A. (2002). Probabilistic basis for 200 sac federal emergency management agency steel moment frame guidelines. *ASCE J. Struct. Eng.* 128(4), 526–533.

Fajfar, P. (1999). Capacity spectrum method based on inelastic demand spectra. In *Earthquake Engineering and Structural Dynamics.* Vol. 28, pp. 979–993.

Fajfar, P. and Gasperic, P. (1996). The N2 method for the seismic damage analysis of RC buildings. *Earthquake Engineering and Structural Dynamics.* 28(1), 31–46.

Fajfar, P. and Krawinkler, H., eds (1997). *Seismic Design Methodologies for the Next Generation of Codes.* Balkema Publishers, Rotterdam.

FEMA-350 (2000). *Recommended Seismic Design Criteria for New Steel Moment-Frame Buildings.* Developed by the SAC Joint Venture for the Federal Emergency Management Agency, Washington, DC.

FEMA-356 (2000). *Prestandard and Commentary for the Seismic Rehabilitation of Buildings.* Federal Emergency Management Agency, Washington, DC.

Freeman, S.A. (1978). Prediction of response of concrete buildings to severe earthquake motion. *Douglas McHenry International Symposium on Concrete and Concrete Structures*, ACI SP-55. American Concrete Institute, Detroit MI. pp. 589–605.

Gupta, B. and Kunnath, S.K. (2000). Adaptive spectra-based pushover procedure for seismic evaluation of structures. *Earthquake Spectra* 16(2), 367–392.

Haukaas, T. and Der Kiureghian, A. (2003). Finite element reliability and sensitivity analysis in performance-based engineering. *Proceedings*, ASCE Structures Congress, Seattle, WA.

IBC (2000). *International Building Code*. International Code Council, ICBO, Whittier, CA.

Krawinkler, H. (1997). Research issues in performance based seismic engineering. In *Seismic Design Methodologies for the Next Generation of Codes* (P. Fajfar and H. Krawinkler, eds). Balkema Publishers, Rotterdam.

Krawinkler, H. and Miranda, E. (2004). Performance-based earthquake engineering. In *Earthquake Engineering: From Engineering Seismology to Performance-Based Engineering* (Y. Bozorgnia and V.V. Bertero, eds). CRC Press, Boca Raton, FL.

Kunnath, S.K. (2004). IDASS: inelastic dynamic analysis of structural systems. http://cee.engr.ucdavis.edu/faculty/kunnath/idass.htm.

Kunnath, S.K. (2004). Identification of modal combinations for nonlinear static analysis of building structures. *Comput. Aided Civil Infrastruct. Eng.* 19, 282–295.

Kunnath, S.K., Mander, J.B., and Lee, F. (1997). Parameter identification for degrading and pinched hysteretic structural concrete systems. *Eng. Struct.* 19(3), 224–232.

Kunnath, S.K., Reinhorn, A.M., and Lobo, R.F. (1992). *IDARC Version 3.0 — A Program for Inelastic Damage Analysis of RC Structures*. Technical Report NCEER-92-0022, National Center for Earthquake Engineering Research, SUNY, Buffalo, NY.

Luco, N. and Cornell, A. (1998). Effects of random connection fractures on the demands and reliability for a 3-story pre-Northridge SMRF structure. *Proceedings of the 6th National Conference on Earthquake Engineering*, Seattle, WA.

Mahaney, J.A., Paret, T.F., Kehoe, B.E., and Freeman, S. (1993). The capacity spectrum method for evaluating structural response during the Loma Prieta earthquake. *Proceedings of the National Earthquake Conference*, Memphis, TN.

May, P.J. (2002). *Barriers to Adoption and Implementation of PBEE Innovations*. Technical Report PEER 2002/20, Pacific Earthquake Engineering Research Center, University of California, Berkeley, CA.

Miranda, E. and Aslani, H. (2003). *Probabilistic Response Assessment for Building Specific Loss Estimation*. Report PEER 2003/03, Pacific Earthquake Engineering Research Center, University of California, Berkeley, CA.

Moehle, J.P. (1992) Displacement-based design of RC structures subjected to earthquakes. *EERI Spectra*, 8(3), 403–428.

Moehle, J.P. (1996). Displacement based seismic design criteria. *Proceedings of the 11th World Conference on Earthquake Engineering*, Acapulco, Mexico.

OpenSees (2003). Open system for earthquake engineering simulation. http://opensees.berkeley.edu.

Park, R. and Paulay, T. (1976). *Reinforced Concrete Structures*. John Wiley & Sons, New York.

Porter, K.A. (2002). An overview of PEER's performance-based earthquake engineering methodology. *Proceedings, Conference on Applications of Statistics and Probability in Civil Engineering (ICASP9)*, July 6–9, 2003, San Francisco, CA.

Priestley, M.J.N. and Calvi, G.M. (1997). Concepts and procedures for direct-displacement based design. In *Seismic Design Methodologies for the Next Generation of Codes* (P. Fajfar and H. Krawinkler eds). Balkema Publishers, Rotterdam, pp. 171–181.

Reinhorn, A.M. (1997). Inelastic analysis techniques in seismic evaluations. In *Seismic Design Methodologies for the Next Generation of Codes* (P. Fajfar and H. Krawinkler, eds). Balkema Publishers, Rotterdam. pp. 277–287.

SEAOC (1995). *Vision 2000: Performance Based Seismic Engineering of Buildings*. Structural Engineers Association of California (SEAOC), Sacramento, CA.

Sivaselvan, M.V. and Reinhorn, A.M. (2000). Hysteretic models for deteriorating inelastic structures. *J. Eng. Mech.* 126(6), 633–640.

Uang, C.-M. (1991). Establishing R (or R_w) and C_d factors for building seismic provisions. *ASCE J. Struct. Eng.* 117(1), 19–28.

Uang, C.-M. and Bertero, V.V. (1991). UBC seismic serviceability regulations: critical review. *ASCE J. Struct. Eng.* 117(7), 2055–2068.

Umemura, H. and Takizawa, H. (1982). *Dynamic Response of Reinforced Concrete Buildings.* Structural Engineering Documents 2, International Association for Bridge and Structural Engineering (IABSE), Switzerland.

Vamvatsikos, D. and Cornell, C.A. (2002). Incremental dynamic analysis. *Earthquake Engineering and Structural Dynamics* 31(3), 491–514.

Index